THE FABRIC OF THE UNIVERSE

A Universe of Motion

RIEN VAN MOOK

For my lovely wife RIAN
Because she felt that this had to be written for our grandchildren
May many others also enjoy and gain knowledge from this

Copyright 2017 by Rien van Mook

Revision and Update 06-2020
rienvanmook@hotmail.com

This writer's earlier books (in Dutch)
De Wording van de Mensheid   Edition Boekscout Soest
De Structuur van het Universum   Edition Boekscout Soest   Now out-of-print

All Rights Reserved. No part of this publication may be reproduced, stored in a computerised data file, or published in any form or by any means, electronic, photocopy, recording, or by any other way, without prior written consent of the author.

Disclaimer
The information given below is not necessarily correct, error-free or complete. Therefore, no rights can be derived from the information given in this book.

Table of Contents

Preface — 5

Chapter 1 Physics — 8
1.1 Basic Principles
1.1.1 Motion
1.1.2 Scalar Motion
1.1.3 Multi-dimensional Motion
1.1.4 Structure
1.2 Basic Concepts — 18
1.3 Basic Motions — 20
1.3.1 General
1.3.2 The Progression
1.3.3 Gravitation
1.4 The Model — 25
1.4.1 Space/Time
1.4.2 Time
1.4.3 Forces
1.4.4 Radiation
1.5 Summary and Conclusions — 32

Chapter 2 Quantum World — 34
2.1 Vibrations
2.1.1 Linear Scalar Vibrations
2.1.2 Rotational Scalar Vibrations
2.1.3 Oscillatory Scalar Rotating Vibrations
2.1.4 Effects
2.2 Light — 40
2.2.1 Essence
2.2.2 Spectrum
2.2.3 Summary and Conclusions
2.3 Electromagnetism — 46
2.3.1 Essence
2.3.2 Electricity
2.3.3 Magnetism
2.3.4 Summary and Conclusions
2.4 Gravitation — 57
2.4.1 Essence
2.4.2 Effects
2.4.3 Summary and Conclusions
2.5 Atoms — 64
2.5.1 Essence

2.5.2 String Theory
2.5.3 Summary and Conclusions
2.6 Chemical Elements   72
2.6.1 Essence
2.6.2 Summary and Conclusions
2.7 Molecules   79
2.7.1 Essence
2.7.2 Bonding
2.7.3 Form
2.7.4 Properties
2.7.5 Summary and Conclusions

Chapter 3 Cosmology   88
3.1 The Universe   88
3.1.1 General
3.1.2 Dark Energy
3.1.3 Big Bang
3.1.4 Cosmic Radiation
3.1.5 The Nature of the Universe
3.1.6 Summary and Conclusions
3.2 The Material Sector   100
3.2.1 General
3.2.2 Star Formation
3.2.3 Why Stars Radiate
3.2.4 Star Cluster Formation
3.2.5 Summary and Conclusions
3.3 Galaxies   106
3.3.1 General
3.3.2 Dark Matter
3.3.3 The Electric Universe
3.3.4 The Einstein-Cartan Effect
3.3.5 Summary and Conclusions
3.4 Novae, Pulsars and Quasars   121
3.4.1 Novae
3.4.2 Pulsars
3.4.3 Quasars
3.4.4 Fast Radio Bursts
3.4.5 Summary and Conclusions
3.5 The Genesis of the Universe   137
3.5.1 The Cyclic Universe
3.5.2 Physical Constants

Chapter 4 Epilogue   141

Preface

The only thing that interferes with my learning
Is my education
Einstein

The title of this book has been given as 'The Fabric of the Universe' with the subtitle 'A Universe of Motion'. That implies a complete different explanation for the structure of the Universe, quite different from what current science has to tell us and what is taught in schools and universities.
The laws of nature as found by science are not invalidated in this theory and are completely in harmony with the results of this theory. Only the interpretation of these laws is clearly different from established scientific theories. In the text is in italics referred to some of the most important scientists who have contributed to where science now stands.
This book is based on the work of Dewey B Larson and his Reciprocal System of physical theory (RS-theory), first published in 1959. It is impossible in the short scope of this book to describe the entire RS-theory. Therefore, only the most important aspects are cited that need no long explanations while parts are treated only in bird's-eye-view. In addition, essential aspects will be repeated regularly because the theory is not always easy, while many aspects are not in accordance with conventional scientific thinking. But the most important thing is that on the whole a personal interpretation is given.
The theory given here does not assume a Universe based on matter such as current science assumes, but a Universe based on motion. It reveals 2 fundamental aspects of the fabric of the Universe, namely Space and Time. Space and Time as defined in the context here are therefore written with a capital letter.
Space and Time are 3-dimensional and have an inverse relationship, meaning they have an opposite or reciprocal effect. Space and Time in a reciprocal relationship, expressed in units

such as 'meter' and 'second', means meter-per-second and so speed. In generic form motion.

The Universe consists only of motion. That can take on many forms such as motion-on-motion and this in linear, rotating and oscillating manner and in 1, 2 or 3 dimensions. It is a continuous changing ratio of Space and Time. Because of the high speeds involved we will in the following also refer to such motions as vibration.

In essence there is only motion, vibrations that in their many manifestations the Universe as we know it form. There is no vibration in 'something', there is no 'ether' and there are no 'forces'. Only vibration as such. Vibrations are the basic phenomena that form the Universe. Photons, atoms, forces, energy and radiation are all manifestations of vibration.

That is further explained and detailed in the theory presented hereafter.

Science believes matter to be the most fundamental unit, not vibration. This is clear for example in string theory where infinitesimal small strings are vibrating.

In order not to stray too far from usual conceptions I therefore do use terms used by science such as 'particle' and 'atom' when it should be read as 'unit of vibration'.

On this basis, it becomes possible to explain the fabric of the Universe and all its perceived components, both the small world of atoms and the big world of galaxy systems, and also to connect the small to the big. Everything then comes together as a Jig-Saw puzzle and, in order to understand the Universe, there are no complicated mathematical exercises such as string theory necessary. Where science struggles with many unexplained issues and have lots of loose ends in their theories, the Universe can generally be understood with the theory given hereafter.

There is no pretence that this explanation reflects the ultimate reality as truth, but it shows that what is taught at high schools and universities is not necessarily the best explanation for the world and the phenomena around us. Ultimately, as science teaches us, this theory is based on theoretical models of reality and therefore only an approximation of reality.

In this book we will explore the implications of the theory in the quantum world and the cosmological world.

My hope is that this book offers an example for students to develop a personal way of thinking, because scientists are too often caught in their own speciality, collegian pressure and financial interests to deviate from the established path.
It will require out-of-the-box thinking, in deviation from the tunnel view induced by indoctrination resulting from upbringing, education, religion, science and the media. Note in this respect for instance that new technology patents are currently reviewed for acceptance according to "known laws of science". How will that bring us forward?
An example of such new way of thinking and understanding, in the theory of this book, is the clarification of the nature and properties of general scalar motion in the Universe,.The existence of this type of motion, which has magnitude only and no inherent direction, turns out to be a primary physical phenomenon. As will become clear distributed scalar motion is for instance responsible for an effect that science sees as "force".
The essence of Space and Time does not belong only to the discipline of physics, or cosmology, or philosophy. Hence, with an open mind an individual can work towards own truth finding, because truth does not come from authority, and intelligence is not demonstrated by perfectly copying the socially and scientifically accepted vision. And brilliant ideas have never been devised in teamwork.
Hopefully, this book will contribute to realize my hope, specifically for my grandchildren, to develop their own independent way of thinking.

Education is the kindling of a flame
Not the filling of a vessel
Socrates

Chapter 1   PHYSICS

That's the most exciting idea I know
That nature is much simpler than it looks
Steven Weinberg

1.1   Basic Principles

1.1.1 Motion

Science sees the world as matter in space and sees time as the canvas on which the history of the Universe is being projected. Matter is the basic unit in this and that appearing in space. That immediately calls the question 'If matter is so basic what then is matter?'
It should be clear that the outcome of scientific experiments can be explained in several ways and that we only create models to get insight into reality.
The theory given here is a different model than the general model advocated by science and thus gives a different interpretation than science. However, in its relative simplicity it is seen as better able to model reality (Occam's razor).
The theory given here is based on the fact that there is a unit that is much more fundamental than matter, namely motion. The world only arises from motion. Not motion of something. No, motion itself is the most fundamental feature. There is nothing else! Through motion everything is formed, also matter.
And this motion is the continuous change of Space and Time. This is a process; and a process leads to a product, for example matter.
The theory given here replaces a cosmos that exists in space and is clocked by time, by a Universe consisting only of motion of Space and Time. The terms cosmos and Universe in the above statement are seen as different and the terms will be clarified later.

According to the theory of this book the Universe consists of units of Space and units of Time. They consist of discrete uniform units and fractional units are not possible.
Space (S) and Time (T) do not exist in themselves but it is a relationship. S/T or T/S.

Space/Time   or   Time/Space

Space and Time are inversely proportional, equal but opposites. Space changes to Time and Time changes to Space in an eternal dance.

> *The more Space the less Time.*
> *And the less Space the more Time.*

And because Space is 3-dimensional to us, Time must also be 3-dimensional in this relation. So we should not see Space in the sense of a medium! And the same goes for Time!
Space/Time is therefore variable and thus dependent on where you are in this Universe!
The motion is, by definition, a ratio of Space and Time and that ratio is inversely proportional. That is, we take a unit of Space and divide it by a unit of Time, for example meters per second, m/sec. And that is velocity! And that velocity is variable depending on the Space/Time ratio. In this theory that is the general term Motion. The base rate in the Universe is seen as the change of 1 unit Space per 1 unit Time in 3 dimensions. It is the mode of rest of the Universe. The theory in this book sees this base rate as the speed of light. Therefore, the speed of light is also 3-dimensional. And that means the speed of light over each of the 3 dimensions.

1.1.2 Scalar Motion

Each unit of Space or Time in the relationship S/T changes in a scalar mode, which means that the units have a magnitude only and no specific direction. A scalar motion means by definition that there is no specific direction but only a value, a magnitude. The

value of units of both Space and Time in the relation changes continuously, affecting the scalar value of the motion.
We are used to vector system motion. This is motion relative to a fixed reference system. It has a magnitude and a direction in a reference system. The effect of vector system motion thus depends on direction and magnitude of the motion.
Scalar motion on the other hand has magnitude only. In comparison with a vector system a scalar system has distinctly different effects. This because setting a reference system will result in a different geometry with different effects. In such a reference system a scalar motion will appear as an acceleration (positive or negative).

In example:
Take a line: Y--------------------------X--------------------------Z

If X is a point on a line YZ, and X moves towards Y, then will in a vector system the distance XY become shorter and the distance XZ will lengthen.
However, in scalar motion both XY and XZ will grow longer, or both will shrink!

Although scalar means that a change occurs in all directions simultaneously, scalar motion can be positive or negative.
When Space increases with Time the motion is called positive, i.e. outward. A negative motion which decreases with Time is thus inward.
To effect this the reversal of direction must apply to only one component, Space or Time. Reversal of both components would obviously not alter the direction.
The motion in Space is an expansive outward scalar motion like a balloon that expands or a cake that rises, and this at every point within the balloon or cake! The opposite occurs with a motion in Time, which is then an inward shrinking scalar motion as a balloon that depletes. Both motions occur simultaneously in different ratios affecting all units of Space and Time in the Universe.
Note that in this definition of scalar motion there is no reference frame! A reference frame can be seen as setting an immovable

point somewhere. But this fixed point in a scalar environment also moves. By fixing a point of reference we will therefore create vectors.

In example;
When in a vector motion system we put vectors of same magnitude in a point A in all possible directions, the effect will be no motion. The vectors will cancel out and point A will not move. In similar scalar motion however A will still move, outward or inward.

We use reference points in our thinking and then set an axis system for our measurements. However, the Universe cannot be put in a reference frame.
To illustrate this we use our example of the line with 3 points Y-X-Z again. In a vector motion system the whole system and its motion can be put in a reference frame and will stay the same if we move it to any place within the reference frame (translational motion). But in a scalar system only one point on the line can be identified as fixed in the reference system. The other 2 points will move in a translational motion. Such motion will then appear as positive or negative acceleration in a vector motion system. (Negative meaning deceleration).
Science has thus made the mistake of setting a reference frame in the Universe, measured speeds of galaxies and found the Universe expanding (*Hubble*). The measured distances were then set as vectors and subsequently traced back to a central point. As a result science has incorrectly concluded that the Universe must have had a beginning, the Big Bang.

No direction in scalar is the equivalent of infinitely many directions, which is the infinity of zero! This explains why scientists in their mathematical calculations repeatedly get stuck in an infinite number. These calculations are based on vectors, on direction.
A relevant example is the famous *Michelson-Morley* experiment, which demonstrates that the speed of light in all directions has the same constant value. That came as a surprise to science. In the experiment a reference frame was set and science expected

different results for the speed of light in different directions. It did not recognize the scalar motion.

Also *Einstein* in his theory of relativity based this on vector measurements. *Einstein* thus deducted that Space must be curved. And therefore ran into problems with paradoxes and measurement issues of time and length. He did not recognize that the Universe consists of distributed scalar motion as we do in the theory of this book!

The scalar feature of motion in 3 dimensions causes the effects to remain Euclidean. There are then no effects like curved space and such as *Einstein* thought.

Absolute time and/or absolute length measurements in the Universe are not possible, as Time and Space are interdependent scalar concepts.

On basis of the 3-dimensional aspects of S/T, scalar motion in our theory can take place in more than one dimension. Scalar motion can take place only in uniform discrete units in each dimension.

The natural reference level for scalar motion is one unit Space per one unit Time, which is in this theory defined as the speed of light. In a vector motion system the condition of rest is zero; in a scalar system its magnitude is thus one. One unit Space over one unit Time, or the inverse. The minimum speed is thus 1 for both normal and inverse speed.

These multi-dimensional distributed scalar motions will appear as the effects of the fundamental forces and constants of science, as we will see later (chapter 1.4.3 and chapter 3.5.2 respectively).

1.1.3 Multi-dimensional Motion

Scalar motion can take place in three scalar dimensions. However, only one dimension can be represented in a spatial reference system. The other motions cannot be observed. This is an important point, that we will encounter later when we address aspects that arise as a result of motion in the unseen dimensions. *Only one dimension of vibration is manifest in our material world, while the other 2 also affect our world.*

In scientific measurements of the speed of light we measure the maximum velocity in a vacuum as constant. However, this measurement depends on the reference frame in which that measurement is made. In a 3-dimensional Space/Time however, as per our theory, the light speed can be exceeded several times. Each of the dimensions stands alone, and in each dimension will the maximum speed be the speed of light. The maximum scalar light speed is therefore 3 times the speed of light.

This finding that light speed is the maximum speed according to the current physical insights is because the measurement is done in a cyclotron where elemental particles (units of vibration) are brought to light speed by means of electromagnetism. That electromagnetic aspect automatically sets the limit! It is the limit in 1 dimension of the 3-dimensional light speed, as we will see in the next Chapter 2 on the Quantum World.

This therefore does not prove that it is not possible to accelerate physical objects to speeds in excess of the speed of light, but it only proves that it is impossible to do so by electrical means.

And because of the 1-dimensional character of electricity while the speed of light is 3-dimensional, the speed of light can be exceeded several times. Although not by electromagnetic means, it is possible by explosive means for example. Not as a chemical explosion that we are familiar with, but in other ways, as will become clear later. This is important and will for instance be made clear in Chapter 3 on Cosmology, where for example, the explosion of a star as nova or a galaxy as quasar will be explained.

1.1.4 Structure

The base rate in the Universe is seen as the motion of 1 unit Space per 1 unit Time in 3 dimensions. That is called unity and is in our theory defined as light speed. It is the state of the lowest energy.

It can be deduced therefore that the Universe consists of two sectors, namely speeds below unity, below light speed, and speeds above unity, above light speed. Both sectors exist within

each other, and are thus defined on basis of the speed of light in each sector.
These sectors are called:

> **Material sector**, where Space/Time is less than 1.
> A speed from zero to the speed of light is the speed in the Material sector.
> S/T = 1/n, where n is an integer from 1 to infinity.
>
> **Cosmic sector** where Space/Time is greater than 1
> A speed of light to infinity is the speed in the Cosmic sector.
> T/S = 1/n, where n is an integer from 1 to infinity.

The term Cosmic sector has been chosen instead of non-Material sector because certain non-Material entities also exist in the Material sector, such as neutrinos.
What science sees as the cosmos is thus only half of what exists in the Universe. The other half is a mirror image with inverted effects.
The Material sector is the world that we are directly aware of, the world that we experience directly. As such, the size of the cosmos is doubled, with a cosmos and an anti-cosmos, which we will call together the Universe.
The term cosmos, as often used in literature, is here seen as equivalent to the Material sector. The anti-cosmos is seen as equivalent to the Cosmic sector.

Because Space and Time are inversely proportional aspects, the Universe looks in the Cosmic sector exactly the same as in the Material sector, only everything works reversed. Matter, for example, moves outward and spreads out in Time in the Cosmic sector. In Time therefore, it forms stars, globular clusters and galaxies, just as in the material sector, but these are concentrated in Time (spread out) and not in Space (clustered).
So, together with the well-known cosmos (the Material sector), there is a mirror image cosmos (the Cosmic sector) in which everything works in reverse.

In the Material sector, Time flows
In the Cosmic sector, Space flows

These sectors do not exist side by side but are intertwined and have opposite properties. In the Cosmic sector, for example, stars consist of atoms that are widely distributed in Space. The temperature of these stars is the opposite and its atoms are therefore cold. Time in the Cosmic sector also has an opposite direction and goes from future to past!
In order to be able to visualize this Cosmic sector one can make a comparison with the Material sector. One dimension of the Cosmic sector will consist of a 3-dimensional pattern of Time coordinates, in which changes of position in Time will take place during the continuous outward expansion of Space. The expansion of Space would then be measured by something analogous to a clock.

The Cosmic sector is an unseen realm that harmonizes with mystical teachings. This makes the Material sector an open system that can communicate with the Cosmic sector. Such open system can explain many features that are otherwise not explainable.
This view differs from science which advocates existence of a Material sector only, resulting in a mechanical, materialistic and closed system model.
In the foregoing definition we find a comparison with the ancient Chinese Yin and Yang. Yin represents the Material sector and is dark, Yang represents the Cosmic sector and is light. They form an inseparable whole and are intertwined. It is at the same time the ultimate polarity. They are the Space and Time of this alternative physics.
Yin and Yang are ancient Chinese concepts that, like Space and Time, are present in everything. The continuous exchange between these two give life to the world, according to these ancient Chinese concepts. These concepts form a whole, equivalent but opposite. These are symbolized in the Tai Chi symbol, the way of the universal life force (chi).
Here we find a rediscovery of ancient oriental wisdom in the physics of this book as explained further below.

Space/Time = 1/1 is the base unit and is the light speed (about 300,000 km / sec, but then in only 1 dimension!)
The light speed forms the boundary between the Material sector and the Cosmic sector and forms the background where everything that is happening is projected, the natural frame of reference for both sectors.
In the Material sector we see velocity, in the Cosmic sector this is equivalent to energy. Why this is so is explained separately hereunder.

A unit of Time is thus equivalent to Energy
And *a unit of Time is also equivalent to Information*

3D Time = Energy = Information

In the Material sector, we see speed
In the Cosmic sector we see energy

Why Unit Time = Unit Energy

m = meter; sec = second; M = mass; V = velocity; E = energy.

Change as Space divided by Time is speed, so m/sec.
In the reverse situation, change as Time divided by Space must be sec/m.

Motion in 1 dimension is m/sec.
Motion in 2 dimensions m^2/sec^2 and in 3 dimensions m^3/sec^3.

Based on the inversely proportional situation (in Time), the reverse of motion in 3 dimensions is resistance to motion (so sec^3/m^3).
Resistance in 3 dimensions is inertia, mass inertia, or mass.

The formula for Energy is $E = MV^2$ (so mass x speed x speed).
Mass multiplied by speed: sec^3/m^3 x m/sec = sec^2/m^2 (moment)
And with another multiplication sec^2 m^2 x m/sec = sec/m.
        And that is energy

Time divided by Space is thus Energy and the reverse of speed (sec/m).

Thus, in the Material sector, we see speed.
In the Cosmic sector we see reverse speed, is energy.

1.2 Basic Concepts

The following concepts now form the basis for our understanding of the world:

- Space and Time are both 3-dimensional, homogeneous and isotropic (= equal everywhere) and both consist of discrete units. Hence, no 3-dimensional space with a 1-dimensional time as *Einstein* proposes.

- The world can be understood with Euclidean = flat geometry. Hence, no Space/Time curvature as *Einstein* argued. That certainly makes for simpler mathematics.

- Throughout the Universe the relationship between Space and Time is scalar (= magnitude only, no direction), inversely proportional and isotropic (= distributed evenly throughout). This also differs from science which uses as base vectors = direction.

That leads to:

- Mutuality.
  Space and Time are inversely proportional. An increase in Space is equivalent to a decrease in Time and vice versa.

- Quantification.
  Space and Time come in discrete units.

- Symmetry.
  Space and Time have the same characteristics. Both have 3 dimensions and propagate in the same way. This gives harmony, balance and beauty.

- Economy.
  A huge variety of effects arise from change in only two elements, Space and Time.

On the basis of these concepts, reality can be explained without having to use complex mathematical formulas.
This applies to the small, the quantum world and also to the big, the Universe, as well as to the connection between these two worlds, the light speed.
The properties of Space/Time that apply to the quantum world also apply to the structure of the largest galaxy.

As the saying goes: "***As above, so below***".

1.3 Basic Motions

1.3.1 General

In scientific way of thinking, the 3-dimensional Space is fixed and the 1-dimensional Time moves forward. In our view however they are both 3-dimensional and move both. This motion is a Space/Time relationship and is an outward motion in Space and an inward motion in Time. We are used to motion in Space, but we must realize that motion in Time is also possible. (This does not mean time travel, but motion of Time itself).
That motion consists of discrete minimum units, units of Space and units of Time. A quantum Space and a quantum Time, and then in different proportions. There are no fractional units, such as half a unit. (That will for example be an issue when we discuss FRBs in Cosmology later).
The unit motion of 1 unit Space per 1 unit Time is the basic motion, the base speed, the speed of light in our definition. We will hereafter call it the Progression.
On this base motion other motions may be possible. Think of the base motion as a carrier wave, as in telecommunications, on which other vibrations may be modulated to convey information. These other motions will also have the effect of scalar 3-dimensional vibrations, and are responsible for all material and non-material objects in our world. This will be further clarified in Chapter 2 on the Quantum World.

Space and Time exist as an inversely proportional relationship. As each point in Space moves outward from all other points, so move all points inward towards each other in Time.
What is an effect on the Space side is the opposite on the Time side. It is Space divided by Time and vice versa.

> **The more Space, the less Time.**
> **And the more Time, the less Space.**

In *Einstein's* relativity theory for example, this is expressed such that when velocity is nearing the speed of light *time expands* and *length shrinks.* He referred to this as a "rubber yardstick. Moving rods must change their length, moving clocks must change their rhythm".

And whereas Space is 3-dimensional as we know it, also Time must be 3-dimensional, because Time is the reciprocal of Space. In our world we see only one dimension of Time and do not realize much of the existence of the other 2 dimensions.

1.3.2 The Progression

The above basic motion of 1 unit of Space (quantum) per 1 unit of Time (quantum) is the condition of rest in the Universe and forms the basis from where all activity begins.
It is light speed in 3-dimensions.
This general scalar motion will be called the **Progression** from here on.
The Progression is thus the basic scalar motion of the fabric of Space/Time. That is why *Einstein* went wrong when developing his theory of relativity. When he envisaged himself sitting on a beam of light he missed the scalar effect of the Progression and had to develop space distorting effects to compensate.

The motion of the Progression is outward in Space and inward in Time. It is the motion of the vacuum.
In the Cosmic sector the scalar motion of the Progression is directed inverse and thus opposite to the motion in the Material sector, hence inwards.
From the neutral condition of the base speed, all S/T quanta expand with the speed of light away from each other. So, both Space and Time move at a reference level of one unit speed.
And unit speed is equivalent with unit energy.
In our model we see this basic speed (the Progression) as the speed of light. That is the natural reference value in 3 dimensions.
To visualize: With the Progression the *Now* moves forward in the same manner as the *here.*
Meaning that at light speed there exists a permanent here and now.

The light speed unity should be seen as a domain. A domain of perfect uniformity where nothing happens and nothing can happen. It is scalar so has a value (200.000 km/sec) and motion in all possible directions.
It is the "body" of the Universe, the essence.
In illustration: As such a photon can travel from one end of the Universe in an instant in our view, because it is already everywhere.
As science would say, it is non-local. See chapter 2.2.1 for further clarification.

In practice the scalar movement of the Progression means that an object will move with the Progression when not acted upon by another outside agency. But when we measure the speed of a galaxy from our vantage point on Earth, our measurement will be affected by such outside agency, namely gravitation. That implies that motion of matter affected by gravitation will deviate from motion with the Progression. We will discuss this effect later in relation to the view of science that the cosmos expands on basis of their measurements.
Hence, this clarifies why objects that are not affected by gravitation, (without mass, such as a photon or an electron) travel with the Progression, and as such are part of the permanent here and now.

To make something happen, there must be a deviation from the Progression, from the base speed either in Space or in Time. Deviations from this standard velocity arise if there will be more Time units per Space unit or more Space units per Time unit. Such deviations on this base motion will occur locally. The scalar property enables a change of direction, a local reversal of speed. It does not change the original scalar direction, because scalar means no specific direction but all directions simultaneously.
This will happen as a linear vibration opposite to the outgoing or ingoing direction of the Progression. On such linear vibration other modes of vibration are possible such as rotational vibrations and oscillatory vibrations. This because a rotating vibration by itself is

not possible, as there is nothing to rotate in the neutral condition.
We will return to this in detail in chapter 2.1.
These vibrating motions will create the world as we know it.
It results in the creation of a sector where general speeds are less than the speed of light (which we have called the **Material sector**), and a sector where general speeds are greater than the speed of light (which we have called the **Cosmic sector**). As a whole, these sectors form an outgoing (in Space) and ingoing (in Time) 3-dimensional scalar motion, the Universe.

1.3.3 Gravitation

A scalar motion on the Progression which has an opposite effect to the Progression is a move towards Time, hence in Space inward and in Time outward. This can happen in all 3 dimensions, in all the three modes of vibration.
The vibration in <u>all three</u> modes we will call the **Gravitation** from here on. It is the motion that produces matter.
As such Gravitation is nominated here as a special basic mode of vibration, as it obviously affects matter in the Universe. We need to specifically mention it here to clarify some issues to do with matter. In Chapter 2 on the Quantum World we will review all 3 possible dimensional modes of vibration, the 1-2 and 3 dimensional modes. There also more detail on the 3-dimensional mode of Gravitation.

We become aware of the 2 specific scalar motions, Progression and Gravitation, by the way matter in the Universe is subject to it. In Space all galaxies move away from each other with the expanding motion of the vacuum, the Progression, while the matter of the galaxy travels inward as a result of Gravitation (a motion opposite to the Progression). This also applies to our own galaxy, the Milky Way.
It is important to recognize that the Progression does not apply to the atoms of galaxies. Only the vacuum of space is subject to it. Galaxies are stuck on the figurative balloon as dots or wrapped-in as crisps in the cake. Each quantum S/T in the vacuum moves away from any other quantum S/T. Matter in the galaxy shows the opposite move, inward through Gravitation.

We therefore see the effect of the 3-dimensional scalar motions of Progression and Gravitation in the manner that matter drifts as a result of both motions. Gravitation makes the motions of the galaxies visible.

Gravitation in Space has a visual effect locally at the point where in scientific view inertia arises. That is the point where an atom forms. In the Cosmic sector, the scalar motion works exactly opposite to the motion in the Material sector. Here the motion of the Progression moves inwards, and Gravitation is thus directed outwards (and thus separates matter).

This is an important aspect that will be encountered many times more in the following.

As stated before we will return in detail on the effects of Gravitation in chapter 2.4.

1.4 The Model

1.4.1 Space/Time

On basis of the definitions of the foregoing chapters we can now develop our model of the Universe.
We see 4 areas in the Universe, each of which has different characteristics, but is characterized by the prevailing speed. This as per the following diagram.

| **Time** | **Time-Space** | **Space-Time** | **Space** |

In these sectors the following applies.

The Material sector: = Time and Time-Space

- Time; 1 unit of Space has many units of Time.

- Time-Space; Units of Time and Space are divided, but the units of Time are greater than the units of Space

The Cosmic sector: = Space-Time and Space

- Space/Time; Units of Space and Time are divided, but the units of Space are greater than the units of Time.

- Space; 1 unit of Time has many units of Space.

The Space to Time ratio increases from left to right in the diagram above. These sectors do not exist side by side but are intertwined and are 3-dimensional!
At the left in the diagram Time is infinite, Space is one unit and speed is zero. Here we find the region popularly called the vacuum energy or zero point energy.
As we will see later, we could for example reduce Space to one unit through torsion and thereby allow an infinite amount of energy to be released. Should this technology be engineered the problems of the world would vanish.

On the boundary of Time and Time-Space, Time is reduced to one unit, Space remains constant and speed is thus unity in one dimension.
Further to the right in the diagram, Space increases in the other dimensions.

At the centre of the chart, on the boundary of Time-Space and Space/Time, the speed is the speed of light in all 3 dimensions, the Progression.

The right side of the diagram is the left-hand duplicate but in reverse. Starting from the middle and starting with the reverse unit of speed, Time becomes smaller and is at the limit of Space/Time and Space just enough to achieve an inverted unit of speed in 1 dimension. At the right in the diagram, Space is infinite, Time is one unit and inverted speed (energy) is zero.

We experience things from the Time-Space area. The left half of the above diagram is what we have called the Material sector of the Universe. The right half is what we have called the Cosmic sector.

If we go to the left, Space becomes smaller and Time increases. Objects become spatially smaller and last longer.
In illustration: This effect can explain how we might be able to travel to the stars. For example, if we want to travel to the stars, we should do that by using the technology of travelling in Time. Then, distances become shorter and more Time becomes available. Thus, our travelling vehicle would for an outside viewer seem to be getting smaller at the location where it is perceived. It would turn into a dot of light and ultimately become invisible.

If we go to the right, Space increases and Time decreases. Objects become spatially bigger and less Time becomes available. Our Time-travelling vehicle would become slowly visible again and larger when viewed from outside.

1.4.2 Time

Time in our awareness moves forward because of a property of the Universe called 'entropy', roughly defined as the level of disorder, the higher the entropy the more chaos.
Entropy (chaos) only increases in the Material sector and there is almost no way to reverse a rise in entropy after it has occurred.
In illustration: It is easy to break a glass, but almost impossible to have the broken pieces rearrange itself into the former glass.
The fact that entropy increases is a matter of logic; there are more disordered arrangements of particles possible than there are ordered arrangements, and so as things change they tend to fall into disarray. This is the arrow of Time.
In our cosmological model the entropy increases in the Material sector only, while in the Cosmic sector the entropy decreases. The arrow of Time is therefore opposite in these sectors.

The Cosmic sector turns chaos into order

To clarify how time travel should be seen in this theory herewith an example.
We can travel on different occasions to a specific coordinate in the Material sector, but then only at different times. You cannot go to Paris on two different occasions and then arrive at the same date/time. And the future is open in this sector.
And in the Cosmic sector exactly the opposite is true. We can travel on several different occasions to a specific coordinate in Time, but then not to the same location in Space. And the past is open in this sector.
It is therefore not possible to travel to a specific location where a specific event took place at exactly the same date/time. Thus, there is no paradox in time travel, such as for example the possibility of being able to kill your ancestor.
Science fiction is usually based on time travel following one-dimensional clock time, to an earlier or later date. That is per the above in principle impossible.
So, what has happened cannot be altered; it has happened.

Application of *Einstein*'s relativity theory without taking into account scalar aspect of Space and Time creates a time paradox. If traveller A leaves the Earth at a speed that approaches the speed of light and returns later, traveller A is then younger than person B that stayed behind on Earth.
However, according to the same theory, it may also be said that the Earth is moving away at high speed and that on returning the person B on Earth is younger than A, who stayed behind. A is then older than B.
That paradox arises because a reference frame is applied and the scalar effect (no directions) is not recognized.

1.4.3 Forces

According to science, bosons represent the 'forces' in nature. Bosons are the carriers of a specific force according to science, and in particular:

- For the electromagnetic force the Photon.
- For the weak nuclear force, the W-boson and Z-boson.
- For the strong nuclear force, the Gluon.
- For gravitation, the Graviton (not yet found).

However, in accordance with our theory there are no forces and therefore bosons do not have the effects that science assigns to it. We will explain these effects in accordance with our theory later for each of these 4 fundamental 'forces'.
For now it suffices to note in illustration that a photon is a linear scalar vibration on the Progression, as will be further clarified in chapter 2.2.1. As such it sits on the Progression and therefore cannot and does not transfer the electromagnetic force.
(The explanation for the electromagnetic effect comes later in chapter 2.3)

In our model the fundamental forces as recognized by science, such as gravity, the electromagnetic force, and the weak and strong nuclear forces, do not exist. Forces in a scalar environment

appear as a geometrical effect from the motions of Space/Time.
The effects of fundamental forces are an inherent property of
scalar motion as evidenced by the fact that these effects have no
specific direction and act as an autonomous entity generated by
the motion.

A scalar motion in its effects will differ from a vector motion,
occurring when a reference frame is introduced. See our earlier
example of the line XYZ. It will now become a distributed scalar
motion and effects will be what science calls 'forces'.

A uniform vector motion does not show a force. Energy has to be
supplied to change the motion. In other words a force is needed to
accelerate a vector motion.

Translatory motion of a scalar entity in a reference frame shows as
an acceleration, hence what in a vector motion system would be
called a force. So, when a reference frame is applied, the motion
shows as accelerated, and we call it a distributed scalar motion.
This shows as the effects of a field. A field not to be viewed in the
sense science views a field. It is the 'force' aspect of a distributed
scalar motion, the quantity of acceleration, with the same relation
to that motion as a force has to a vector motion. Difference is that
a vector system force has a specific direction whereas the force of
the scalar field is directionally distributed.

This will be clarified with an example.

The law of Newton states: **F = m.a**
Force = Mass x Acceleration, where the force is a vector and
therefore has a magnitude and a direction.
In this, acceleration is a deviation from a uniform motion.
But this change in a uniform motion actually works scalar, thus as
a balloon that expands. The effect is seen on a surface. But if the
diameter of a balloon is doubled, the surface will be quadrupled.
This explains the quadratic law for gravitation, with the 'force'
decreasing quadratically with the distance.

When an object approaches light speed its mass increases
according to *Einstein* in his relativity theory. In his view mass will
show a proportional increase with acceleration when nearing the

speed of light, as there is a limit set by science to the speed of light. In the theory of this book another explanation can be given. Following the above law of Newton one might also assume that F (the effect of the force exercised) becomes smaller. And that is what the resulting effect in this case factually is.

Because what really happens is that (as previously explained) transfer of a system in a scalar environment will show as acceleration (in this case negative or deceleration) and thus will show the effect of force in a reference system. It may in a reference system be visualized such that the force operates at an ever larger angle on the mass, as a result of the scalar workings, and thus will have an ever lower effect. And when the light speed is reached, the force will work at 90 degrees on the mass and thus no longer have any effect.

Moreover, according to *Einstein* himself, with his famous equation $E = mc^2$, the amount of energy in the mass would inexplicably increase with the increase of that mass! And where would that energy then come from?

There is therefore no curvature or distortion of Space, as *Einstein* argues in his theory of relativity. It is the effects of distributed scalar motion that shows as acceleration in a reference frame.

The effect is also in line with the law of diminishing returns. This law states that the ratio of incremental output of a general physical process does not remain constant, but eventually decreases and ultimately reaches zero.

1.4.4 Radiation

As clarified in chapter 1.3.2 on the Progression, the radiation of stars is conveyed by photons on the Progression. The motion is thus with the speed of light, and is not affected by gravitation. Radiation is thus to be viewed as the <u>motion</u> of photons on the Progression. Not as the photon itself. This effect and the resulting properties will also be reviewed further in chapter 2.2.

Radiation is not transmitted. The photons remain permanently in the same Space/Time location it originated, but Space/Time itself, the progression, progresses, carrying the photon with it. The photon is thus able to act on any object not carried along with the Progression.

The point we also wish to make here is that radiation from both the Material sector and the Cosmic sector is directly visible. As the speed of light forms the boundary between both the Material sector and the Cosmic sector, and radiation is a property of the Progression, radiation can be detected from both sectors, with brightness depending upon distance.

Radiation of Cosmic objects is thus (to some extent) visible from the Material sector. That radiation comes from objects that in the Cosmic sector are spatially distributed. (A star in the Cosmic sector is spread out in Time). As such the radiation of the Cosmic sector is not localized and we see that cosmic radiation in the Material sector as of low strength and isotropic (= everywhere equal).

That corresponds to the Cosmic Microwave Background radiation (CMB) of 2.7 degrees Kelvin.

Whereas science sees the Cosmic Microwave Background (CMB) radiation as a result of the Big Bang the above provides a different explanation for it. We will return to this in chapter 3.1.4.

1.5 Summary and Conclusions

In summary, and in deviation from current scientific theory, we conclude that:

- Motion as Space/Time modulation is the most fundamental feature of the Universe, not matter. This motion is velocity (m/sec), depending on S/T ratio.

- Space/Time motions propagate scalar and Space/Time is therefore Euclidean and not curved by matter.

- Space and Time motions are both 3-dimensional, in a reciprocal relationship. The more Space, the less Time and vice versa.

- The base rate of velocity in the Universe is seen as the change of 1 unit Space per 1 unit Time in 3 dimensions, and is by definition the speed of light.

- The speed of light forms a boundary between 2 sectors. One sector with general speeds below light speed, called the Material sector. And one with general speeds above light speed, called the Cosmic sector. Both sectors are intertwined.

- Both sectors are identical but with opposite effects. In the Material sector Time flows. In the Cosmic sector Space flows.

- In the Material sector, we see speed. In the Cosmic sector we see its equivalent, energy.

- Radiation from both sectors will to some extent be visible in the Material sector.

- Entropy increases in the Material sector, while in the Cosmic sector entropy decreases. The arrow of Time is therefore opposite in these sectors.

- The base scalar motion of S/T is called the Progression. It is the condition of rest in the Universe and is the basis from where all activity begins. It is the motion of the vacuum. The Progression is light speed in 3-dimensions, outward in Space and inward in Time.

- To make something happen there must be a deviation on the Progression, in the form of more Time units per Space unit or more Space units per Time unit. These will occur locally as a reversal of speed. This can happen as a linear vibration perpendicular to the original motion or a rotational or oscillating vibration on the linear vibration.

- A scalar vibration on the Progression with opposite effect in all three dimensions is called the Gravitation.

- Those principles are applied in a model that defines in essence 4 areas in the Universe with different characteristics, characterized by the prevailing speed.
    Time - Time-Space - Space-Time - Time

- A scalar motion translates as acceleration (positive or negative) when a reference frame is introduced, e.g. by using vector system measurements. This explains the effect of forces as defined by science.

- The Cosmic sector is an unseen realm that harmonizes with mystical teachings. This makes the Material sector an open system that can communicate with the Cosmic sector. This view differs from science which advocates existence of a Material sector only, resulting in a mechanical, closed system model.

Chapter 2   Quantum World

If you want to find the secrets of the universe, think in terms of energy, frequency, and vibration
Nicola Tesla

2.1 Vibrations

In this chapter we will clarify the various modes of vibration. But first a summary of what we have learned.
The continuously changing motion of Space/Time ratios are seen as vibrations. Vibrations are in 3 dimensions and are scalar in all 3 dimensions.
In summary, the light speed propagates scalar in 3 dimensions and consists of 1 unit of Space per 1 unit of Time, called unity. Speeds greater than the light speed form the Cosmic sector and speeds smaller than the light speed constitute the Material sector. This scalar propagation of Space/Time as unity produces the Progression. The Progression goes out in 3 dimensions scalar. Due to the scalar character, locally within this velocity, there can be a reversal in direction. This reverse speed on the original 3-dimensional light speed of the Progression then goes opposite the outgoing direction of the Progression and is as such the basis for other modes of vibration.

We have as such already met the Gravitation, as vibration on the Progression in all 3 dimensions. But other modes in other dimensions on the base motion of the Progression are also possible, and this both outside in Space and inward in Time.
We will review these vibrating modes in this chapter.
Later in chapter 2.6 we will define the specific build-up of vibration modes for each chemical element.

There are in principle 4 different geometric types of vibration modes. These modes of vibration can exist in 1 dimension up to 3 dimensions, affecting both Space and Time. Those are 3 dimensions of vibration (Space/Time), not 3 dimensions of Space!

These vibration modes are:

- Linear scalar vibration (straight outward or inward).
- Linear oscillatory vibration (constant reciprocating vibration).
- Rotating scalar vibration (spirally outward or inward).
- Rotating oscillatory vibration (as in an auto-reversible watch).

Such motions then form, in example, electromagnetic radiation (out with the Progression), in one dimension electricity (out), in two dimensions magnetism (in), in three dimensions matter (in with the Gravitation), etc.
Only one dimension of vibration in Time is manifest in our Material sector. The other two dimensions cannot be represented in our spatial reference framework, though they affect our world too!

Only one dimension of vibration is manifest in our material world, while the other 2 also affect our world.

We will come across this aspect more in this developing story.

2.1.1 Linear Scalar Vibrations

The linear scalar vibration is the basic scalar motion of the Progression, in Space outward, in Time inward.
And equally so is a linear vibration on the base motion of the Progression scalar. In scalar motion (outbound in all directions at the same time) it then forms a sphere of one unit Space per unit of Time. We call this type of vibrating motion a photon, the unity of light.
This mode of vibration shows as spherical (science would say as a 'particle') if it is from our perspective at rest, so at emission and absorption. In other words, when we measure it.
The photon is always at rest because it is a vibrating mode on the Progression, a photon sits on the Progression and as such travels with it.

We see it as radiation from our perspective when we look at it in motion, when we focus on the movement of the Progression. It is then a ray of light.

2.1.2 Rotational Scalar Vibrations

A rotational scalar vibration on a linear scalar vibration on the base vibration of the progression produces a rotating 3-dimensional vibration.
That rotational vibration can be manifest in all 3 dimensions, as in matter, but also in only 1 or 2 dimensions. The rotational speed can also be different in each of the 3 dimensions.
For the rotation around 1 dimension, the rotational pattern is not constant but uniformly changing, from clockwise to anti-clockwise rotation. It is a rotating swing vibration (as in an auto-winding watch) exercised on a linear scalar vibration on the Progression, further clarified in chapter 2.1.3.
For rotations around 2 dimensions, the rotation pattern is constant, but can rotate in two directions, clockwise and anti-clockwise, thus forming a sphere.
Adding a one dimensional oscillating rotation to a 2 dimensional rotating vibration will create an atom. In the Material sector, matter is formed by rotation at a rotational speed that is less than the speed of light. In the Cosmic sector, matter is formed by rotation at a rotational speed that is greater than the speed of light.
Rotational vibrations are at the basis of all matter, from small such as atoms to huge, including planets, stars, galaxies and clusters of galaxies.

Motion as S/T defines velocity only in terms of speed and energy. Not of a particle moving. It is possible to identify particular units of Space and Time independently of any moving object.
A 3-dimensional particle (unit of vibration) in motion encounters translational motion. That is the additional Space that the particle will see as a result of motion. The rotation with 1 unit of Space per unit of Time adds a Space unit. The presence of the additional unit of Space of the rotating particle increases the total amount of Time required to traverse a given amount of linear extension. This

means a lower velocity on the usual basis of measurement. Accordingly, we find that the velocity of light in matter is less than the speed of light.

The greater the Gravitation, the more Time units there are, and the slower Time passes by. Thus, the more Gravitation the slower the measured speed of light.

We will review this in further detail in chapter 2.4.2 on the effects of Gravitation.

As an example the following:
When cooling a gas down to absolute 0 degrees Kelvin all motion of the particles stop. This creates what is called an Einstein-Bose condensate, a cloudy smudge of particles with no defined shape. Scientists have managed to show that light transmitted through such condensate can be made to slow down and even come to an apparent stop.

The explanation for this effect is that additional Space units are created and in the model of chapter 1.4.1 we have moved completely to the right. Here we have created infinite Space and 1 unit of Time.

The view of Space as we have developed it may then be stated as follows:

- For *Newton* Space was a container containing an atom without effect on Space.
- For *Einstein* Space was a container in which an atom would deform Space.
- For us Space is a container containing an atom in which a unity rotation of the atom (= 1 unit of Space per unit of Time) adds a Space unit.

2.1.3 Oscillatory Scalar Rotating Vibrations

A third mode of vibration is an oscillatory motion as in an auto-winding watch.

It is a rotation around 1 dimension. The rotational pattern is not constant but uniformly changing, from clockwise to anti-clockwise.. It is a rotating swing vibration, exercised on a linear scalar vibration or rotational scalar vibration.

Examples of one-dimensional oscillatory vibrations are electrons, positrons, neutrinos and the like. From the oscillatory motion these units obtain what science calls a half-integer spin.

An atom is formed by adding 1-dimensional oscillatory vibrations to 2-dimensional rotational vibrations, as we will see later in chapter 2.6 on chemical elements.

2.1.4 Effects

The possibilities and results of rotational vibrations are as follows and result in the following effects in our Material sector. Keep in mind that the dimensions are dimensions of vibration and not dimensions of space.

- When rotating around 1 axis, sub-atomic particles of the electron group are formed. This rotating and oscillating vibration in one dimension forms an electron. That is a rotating unit of Space. If it is a rotating unit of Time, we call it a positron.
  The rotation pattern here results in the phenomenon of **electric induction**.
  An uncharged electron cannot pass through Space because it is rotating Space. It can therefore go through matter because matter is Time-oriented. We then see that as electric current.

    Effect: Electricity.

- Rotation around 2 axes produces subatomic particles of the neutron group (neutron and neutrino) resulting in the phenomenon of **permanent magnetism**. Electromagnetism is another phenomenon, as we will see in chapter 2.3. Due to the opposite rotational pattern in the 2 dimensions, the

two opposite effects of positive and negative attraction/rejection arise.
Same vibrations reject, different vibrations attract. Therefore, shielding can take place here, this in contrast to the Gravitation effect where shielding is not possible.

Effect: Magnetism.

- The third possibility is created by an added 1-dimensional vibration to a 2-dimensional vibration. This results in what science names mass. This first of all forms the proton, the core of a hydrogen atom. By adding more and more units, and also in combinations, the chemical elements are created.
This is further elaborated in chapter 2.6.

    Effect: Matter.

2.2 Light

2.2.1 Essence

A linear vibration on the scalar basic vibration of the Progression as a result of reversing the direction of Space or Time is also scalar. That forms a linear scalar vibration perpendicular to the Progression.
Such linear scalar vibration on the Progression is called a photon. It is a unit of energy.
Science calls this a particle (the photon) and defines it as a boson, after the Indian physicist *Satyendra Bose*, and finds that it has an integer spin.
For science spin is a bizarre physical quantity. For us it means a full rotational scalar vibration.
This explains the integer spin that science sees in it. This in contrast to the rotating oscillating 1-dimensional vibration that gives a half integer spin.

A photon is thus the simplest vibration on the base vibration of the Progression. A photon is the carrier of energy and forms 'light'.
Because a photon sits on the Progression a photon is essentially everywhere. As science would say, it is non-local.
The photon is a motion on the Progression, hence there is no 'radiation'. It is not a light ray (a vector) since the vibrating motion is scalar.
The unit frequency (one unit of Space per unit of Time) is the Progression, the carrier of photons. The photons in their motion with the Progression are the reason for what we see as electromagnetic radiation at the speed of the Progression, hence at the speed of light.
The frequency of the photon is the Space/Time frequency of the particular rotating vibration on the Progression.

It is known in physics that bosons cannot connect. In our model the explanation for this is that photons are bosons and photons cannot connect because photons are stuck on the Progression and thus is their motion with the Progression.

Why can they show interference effects then, one might ask?
The answer to this is that they interfere with themselves because of their non-local character. Non-local meaning that they are by definition everywhere because they sit on the progression (see also chapter 1.3.2).
This because of the condition that "*the less Time there is the more Space there is*". This means that for a unit of Time the number of units Space is infinite. So, for that particular unit of Time the photon is actually everywhere!

A photon travels with the Progression and thus with the speed of light from our perspective. Thus, a photon cannot transfer force and the speed is always constant.
Nevertheless, we find that the speed of light in matter is reduced. The explanation for this is that the rotational vibration in matter (S/T) creates additional Space (translational motion) where the 'radiation' must pass through. The light must apparently cross more Space and the light speed then becomes less than the as constant measured value. See also chapter 2.1.2.

Because we travel slower than the speed of light and because the photon is carried along with the base vibration of the Progression, and thus with the speed of light, it forms a sinusoidal vibration from our perspective and appears to us as radiation. Science calls this an electromagnetic wave. The Space/Time oscillation is the frequency.
This explains why light is acting as a particle (when at rest, at emission and absorption) and also as a wave motion. It depends on how we look at it!
Note that science is still puzzled by this and cannot explain this. The famous double split experiment demonstrates this behaviour. When a beam of light is directed through a tiny slit, it leaves a patch of light on a sheet behind it, demonstrating streams of particles. With two slits a shows light and dark bands, demonstrating waves of energy. With photon detectors in place in the double slit experiment, again patches of light are found, even when the detectors are used after the experiment, showing again particle behaviour.

The physicist *Richard Feynman* explains this phenomenon by assuming that light does not just follow a straight line, but that light follows all possible paths from a source to an observer. (The Progression in our view).
And if you then add all 'histories' together, most of the possibilities cancel out and only one remains, the straight line (?).

A photon is a unit of energy. Its frequency is a measure of the amount of energy. A photon sits on the boundary of the Cosmic sector, making it the most basic source of energy in the Material sector. And because it sits on the boundary between Material sector and Cosmic sector, Space or Time has no meaning for the photon, and therefore not for light.

2.2.2 Spectrum

The amount of energy in the photon determines the frequency and thereby for example the colour of the photon as we see it. In the rainbow, the colour blue has a higher frequency than the colour red. Colour is an indication of the frequency of vibration and thus is a signal with a variation in Time.
The scientific explanation why we see a colour is that all frequencies of the light shining on a material are absorbed, except for the frequency that is reflected. The reflected frequency gives the colour to the material.
Therefore, for example, we see the sky as blue because the atmosphere scatters the light, and especially the short wavelength and that is blue. That scattered light is present everywhere and that is what we see a blue sky.
In the morning and evening when the sun is low in the sky, the light must traverse more atmosphere and there is thus more scattering of all colours except the long wavelength of red. So we see red reflected in the atmosphere.
It is important to recognize that there is no change in the electromagnetic radiation itself, because photons do not move independently of the Progression. The colour spectrum arises because the world manifests itself in other ways, like a curved mirror gives us a distorted image.

We are in the Material sector and the photons fly us by with the Progression. What is happening between our eyes and the photons is the mirror that distorts.

The electromagnetic spectrum as radiation can be seen as the mark of the boundary between the Material sector and the Cosmic sector of the Universe, as photons exist and remain on that boundary.

The electromagnetic spectrum increases with increasing frequency from radio waves via microwaves to visible light (rainbow colours; from red to blue) and on to X-ray and gamma ray waves.

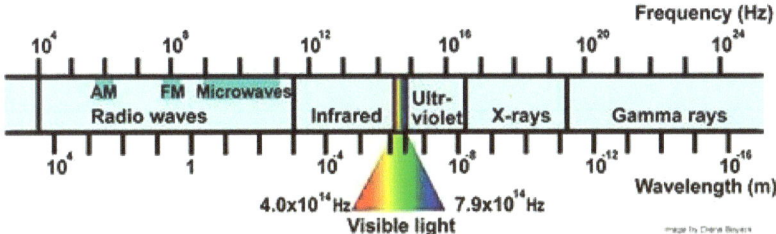

Once more, because the photon is on the boundary of 2 worlds, namely the Material sector and the Cosmic sector, and behaves scalar through 3-dimensions, it appears as a particle or as a wave, depending on how we measure it. It shows itself as a particle if we measure it as it moves with the Progression, at emission and absorption. It shows itself as a wave when we measure it in apparent motion, so when we look at the Progression passing us by.

A good example is the polarization of light, in which the scalar effect is being converted into direction by moving photons through a specific target gap.

Using again the double split experiment with two slits placed at right angles on top of each other, no light will go through the slits. But if we put a third slit between them at a different angle, we see that photons will go through, but now in a relationship that is statistically determined.

In our model this exposes the Cosmic sector.

A test for which science again has no explanation.

In this way we can for example explain the EPR (*Einstein, Podolsky, Rosen*) paradox, namely the quantum entanglement effect. It originated from a thought experiment involving two elemental particles such as, for example, an electron. If such particles with identical spin are paired and subsequently separated, and the spin of one is changed the other particle changes its spin instantaneously to match that of its partner. So for example, if one electron gets an upward spin, the other electron will instantaneously get a downward spin. The distance between these electrons does not matter, they can be lightyears apart. This has been demonstrated recently in real experiments. *Einstein* called this effect "*spooky action at a distance*" because he could not explain how one electron would know what the other does. Our explanation is that the electrons remain connected in Time, which cannot be represented in 3-dimensional Space. If the electrons are intertwined in this way, they take up the same Time unit in the Cosmic sector. In that case, a change in one electron will simultaneously cause an appropriate change in the other electron, just as it would be if they were connected in the Material sector. Contact in Time is subject to the same conditions as contact in Space. This effect is therefore instantaneous and will extend throughout the Universe.

You can also visualize them as the same electron, but then viewed from 2 different sides, projected from the Material sector or from the Cosmic sector.

2.2.3 Summary and Conclusions

In summary, and in deviation from current scientific theory, we conclude that:

- The continuously changing motion of Space/Time ratios are seen as vibrations, basically linear vibrations, rotational vibrations, and oscillatory vibrations, all in three dimensions.

- Difference in dimensions has different effects, as follows:
    1-dimensional: electricity

2-dimensional: magnetism
3-dimensional: matter

- A photon is a linear vibration on the base vibration of the Progression.

- A photon is the most basic unit of energy. The amount of energy in the photon determines the frequency.

- A photon travels with the Progression, with the speed of light. Thus, a photon has no mass and cannot transfer force. Its speed is always constant.

- Photons (bosons) cannot connect to each other, simply because they are stuck on the Progression.

- The velocity of light in matter is less than the speed of light. This because a moving 3-dimensional particle encounters translational motion, which is the additional Space that the particle will see as a result of motion. That increases the total amount of Time required to traverse. This means a lower velocity.

- Because the photon is on the boundary of both sectors, and behaves scalar through 3-dimensions, it appears as a particle or as a wave. The photon shows itself as a single independent unit at emission and absorption, because it is permanently located in the same Space/Time location. A photon is seen as radiation when we measure it as travelling with the Progression.

- Quantum entanglement can be explained because particles remain connected in Time, which cannot be represented in 3-dimensional Space.

2.3 Electromagnetism

2.3.1 Essence

According to science Ionization is the process by which an atom acquires a negative or positive charge by gaining or losing electrons.
But in our theory an atom is an uncharged configuration of vibrations with a net displacement in Time, as further clarified in chapter 2.5. Hence, the atom can get a rotation in Space.
The atom will be called charged when directionally opposed to the Space/Time direction that it modifies. We call it positive when it is Space directed, negative when it is Time directed.
Accumulated charge may be seen as "pressure", directed towards either Space or Time.
Ionization in our model is thus a modification of the fundamental rotation of an atom, towards Space or towards Time. The charge of an atom is thus an element of vibration of the atom and not of electrons that do or do not orbit a nucleus.

Based on the two 1-dimensional rotational possibilities, in Time or in Space, we can explain the difference between charged and uncharged particles. Static electricity and electric current are then 2 different things.
Science sees a proton as a unit with a positive charge and an electron as a unit with a negative charge. However, in view of different masses of the proton and the electron there is then clearly no symmetry.
But in our theory this symmetry is definitively found as follows:

- **A charged electron**
  A charged electron is a rotating unit Time (a quantum energy). That gives a negative charge according to the standard notation of science. As such, electrons can pass through Space. Charged electrons are called static electricity, in Space and in matter.

Lightning is an example, existing of charged electrons (quantums energy) drawn from the zero point.
The following picture shows a plane over London being struck by 3 Lightning strikes on 7 June 2020. With one strike coming from earth. The plane had achieved a positive charge from flying through a storm.

An uncharged electron is a rotating unit Space.
In example:
As charged electrons can pass through Space it is thus convenient to create a vacuum to facilitate it. Hence the vacuum tubes in old radios and TV's. It thus also clarifies why a diode allows current flow in only one direction, from negative cathode to positive anode, and also why heating the cathode assists in functioning of the diode, as it supplies

additional units of energy (rotating units of Time) to the negative pole.

- **A charged positron**
  A charged positron is a rotating unit Space. That is a positive charge. It is the mirror image of an electron.
  An uncharged positron can only pass through Space and not through matter because the rotations are simply incorporated into the vibrating structure of the atom, which also vibrates in Time. Therefore, positrons are relatively rare.
  In example:

  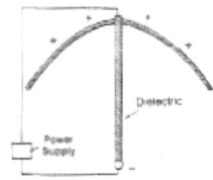

  In a capacitor the negative pole (charged electrons = units of energy) are isolated from the positive pole (charged positrons = units of Space). That produces electrostatic pressure from the negative to the positive pole. A small negative and a large positive pole with mega voltage difference will then produce a propulsion force, and pulsed application can provide directional control. Operation in vacuum conditions would be best, hence in future space vehicles.
  The B2 bomber is said to be already equipped with it. It is called electrokinetics technology.

- **An atom**
  Is an uncharged configuration of vibrations with rotation in Time, so it can also get a rotation in Space, like a positron; that gives it a positive charge. That is what a proton is in our model, not the 'core' of the hydrogen atom as science sees it, but a hydrogen atom with an added vibration.

An electron and a positron thus produce the symmetry in charges, not the proton and the electron as science advocates.
The charge of an atom is thus a pattern of vibrations of the atom and not something that results from mobile charges that do or do not orbit the atomic nucleus.

Charge is therefore not a surplus or shortage of electrons!
In other words: Charges arise from addition of a rotational vibration to an existing rotational vibration of the opposite Space/Time direction.
Matter is Time-oriented and can therefore go through Space, as we accept in our world as normal.

2.3.2 Electricity (one-dimensional)

As clarified before, an electron or positron is a rotational vibration on a linear vibration in one dimension.
An electron that rotates towards Space is an uncharged electron. That electron goes through matter as an electric current because it uses the vibrating potential of direction in Time. The electron does not orbit a nucleus, but goes through the atom as an added rotation to an atomic rotation. But this uncharged electron can then only go through matter, through a conductor, and not through Space.
As such, an electron that goes through matter is the manifestation of electric current. Therefore, <u>uncharged</u> electrons moving through a conductor produce electric current. These are therefore not charged particles, as science already since *Faraday* proclaims, but static electricity in motion!

In other words and as stated before, if the electron rotates towards Space it is an uncharged electron. Therefore, the electron cannot pass through Space because Space to Space is not a vibration. But the uncharged electron may therefore go through matter because matter has a rotation in Time and the relationship of Space/Time is vibration.
So here too, the theory of science has a problem by advocating that electrons pass through the Space between atoms.
The obvious question is then: "Why would they need a conductor"?

A conductor has a certain energy level based on its own temperature. This because the atoms of the conductor are also vibrations. The higher the temperature, the greater the vibration

amplitude. Therefore, a part of the vibration energy of the electrons is required to join the atomic rotation, which gives the effect of 'resistance'. The amount of energy transfer to the conductor is the factor that determines how much current is going through the conductor.
That produces thermal energy, so heat. This addition increases the temperature of the conductor.
Accordingly, a conductor is a perfect conductor, without resistance, at absolute zero degrees temperature.
And although all matter have a net rotation in Time, there are differences. Good conductors have a good net rotation in Time. Bad conductors have part of their rotations in Space in their total rotational pattern. And insulators have little rotation in Time in their overall rotational pattern. Impurities in the conductor also play a role.

Electric current strength is then equivalent to the number of electrons per unit of Time. Electrons are units Space, so electric current is units Space per unit Time. Hence, electric current is motion!
The velocity of the electrons is a measure of 'electric voltage' which then arises between the different endpoints of a conductor. This voltage attempts to achieve equilibrium between the velocities of the electrons at both ends of the conductor.
The thermal equilibrium also plays a role because part of the energy transfer is delivered to the environment. That is why insulation helps.

Different materials have different properties for transmission of the vibration of an electron. This can ensure that when two different materials are in contact, one material passes a certain amount of electrons at a certain velocity, while the other material compensates that by a greater amount of electrons at lower velocity.
An example is the potential difference between zinc and copper, whereby copper as a superior conductor passes more electrons at a lower velocity. If these materials are in contact, this can result in disintegration, in a manner that is called stress corrosion.

Atoms, electrons, neutrons and the like are thus not particles as science proclaims. In our view these are all combinations of different types of vibrations. For example, a neutron will not be taken in by an atom as a neutron, but as a unit of vibration in Time. This action, which is the essence of a neutron, only adds to the existing atomic rotations and becomes an integral part of those vibrations.

Also, electrons and protons do not exist in an atom as individual particles, but form in combinations and together with neutrons, complex forms of vibration. It is only the net direction of vibration, in Time or in Space, which represents the ultimate essence.

Such a complex set of vibrations may then produce a vibrating response in reaction to a stimulus, such as absorption or emission of an electron or a photon.

Or many other forms of vibrations, in example upon destruction by collision in a Cyclotron. More on this in chapter 2.5.1 Atoms.

2.3.3 Magnetism (two-dimensional)

Based upon the possibility of 2-dimensional rotation, we can explain magnetism.

Because there is only a difference in dimension, the aspects of electricity and magnetism are in essence comparable. They exist only at right angles to each other because of the extra dimension. A magnetically charged vibration unit, which has rotating vibration in 2 dimensions, produces 2-dimensional or magnetic effects.

If a 1-dimensional uncharged electron or positron obtains a forward motion, this also forms a 2-dimensional vibration and produces electromagnetism, again a magnetic effect.

Magnetic charges can be positive and negative, but in the Material sector they are positive because rotational vibrations in the Material sector are Time-oriented. Because the rotation viewed from one side looks clockwise and viewed from the other side anti-clockwise, a magnetic 2-dimensional vibration always has 2 poles. Thus, a magnetic charge always has 2 poles, because it are 2 aspects of the same phenomenon.

That also explains why two magnets attract or expel each other, depending on which poles are connected. Like poles reject, different poles attract.
The magnetic effect therefore depends on viewpoint. Thus, a magnetic monopole, like the electric equivalent of the electron, does not exist.
The existence of magnetic monopoles is predicted by various cosmological and scientific theories, but attempts to find them have so far proved to be in vain.
That will probably not change.

Permanent magnets and electromagnetism are also 2 different things.

- Permanent magnets have a 2-dimensional charge due to the dimensions of their rotational vibration. Just like static electricity has a 1-dimensional charge.

- Electromagnetism is a linear inward scalar vibration such as Gravitation, but without one of the 3 rotational vibrations inherent in mass. This due to the addition of the 1-dimensional electric vibration to a conductor.

What happens therefore depends on whether electrons through a conductor neutralize the effect of one of the 3 scalar dimensions of the conductor.
This transforms one dimension of the conductor into a magnet which will appear as a magnetic field around it.
That is the reason why it will produce a shock if a conductor of electric current is touched uninsulated, because a part of the body will be magnetized in one dimension.
Electromagnetism is therefore dynamic. A change will necessitate finding a new balance.

To explain in our model why the electromagnetic 'force' is so much greater than gravitation the following.
Remember what we said earlier in chapter 1.1.3:

"Only one dimension of vibration is manifest in our material world, while the other 2 also affect our world."

The net displacement of the rotations of matter is in Time and that involves rotations in 3 dimensions in Time.
Time is connected to Space as a scalar amount and only 1 dimension of a multi-dimensional vibration in Time can therefore have effect in Space.
And because each of the 3 dimensions can only reach at maximum the speed of light, the entire effect can be transmitted in only one of the three dimensions for each effect.

- A 1-dimensional vibration will produce the electric force.
- A 2-dimensional vibration will produce the effect of 1 divided by the light speed ($1/c$). The magnetic force is thus $1/c$ weaker than the electric force.
- A 3-dimensional vibration will produce the effect of 1 divided by the light speed squared ($1/c^2$). Gravitation is therefore again so much weaker.

This clarifies exactly the how and why electromagnetism and gravitation are linked together. Namely, as a consequence of the number of dimensions in which they exist!
It also clarifies the relation of the strength of their effects.
In scientific language it explains "how electromagnetism and gravity as well as their respective forces relate".
That is the 'Holy Grail' that science is so eagerly looking for!
It is also at the basis of the theory of the electric Universe. See chapter 3.3.3.

As we will discuss in detail in the next chapter 2.4 on Gravitation, a mass A in Space is not 'attracted' by a mass B, but in fact mass A moves to all Space/Time locations of mass B. The same is applicable for the electric and magnetic effects.
Such effects are a vibration of the mass in relation to the fabric of Space/Time and not an action from the one mass to the other, and vice versa.

Therefore, the electromagnetic and gravitational 'forces' are not transmitted by a photon or a graviton respectively, as science advocates. These 'forces' work directly and without a medium, and are a direct result of the vibrating effects on the fabric of Space/Time.
And we have seen earlier that it is not possible to travel in Time twice to a specific place and arrive at the exact same date/time. Something similar we see in the concept of complementarity in quantum physics.

Complementarity means that properties of quantum particles come in pairs, which can only be measured by selecting one aspect of the two. However, the more precise this value is measured, the less accurate the other value can be measured. In the quantum world it is impossible to measure 2 aspects of an object at the same time. (*Heisenberg's* uncertainty principle). For example, if one wants to measure the location of an electron, one cannot accurately determine its speed and vice versa.
Another example is the particle/wave aspect, previously discussed for photons and for the first time experimentally demonstrated for electrons in a double gap experiment in Japan in 1987.
This is a clear indication therefore that such phenomena have a component in the Material sector and in the Cosmic sector! Measurement of location implies measurement of emission/absorption in the Material sector, while measurement of speed implies measurement in the Cosmic sector of the Progression.

2.3.4 Summary and Conclusions

In summary, and in deviation from current scientific theory, we conclude that:

- Atoms, electrons, neutrons and the like are not particles. These are all combinations of different types of vibrations. It is the net direction of vibration, in Time or in Space, which represents the ultimate essence.

- An atom is an uncharged configuration of vibrations with a net displacement in Time.

- An electron is a rotational vibration on a linear vibration on the Progression in one dimension.

- Ionization is a modification of the fundamental rotation of an atom. Positive in Space and negative in Time. The charge of an atom is therefore a pattern of vibrations of the atom and not something that results from mobile charges (electrons) that do or do not orbit the atomic nucleus.

- Electrons are units Space, so electric current is units Space per unit Time.

- Electric current does not consist of electrically charged components, but of uncharged components.

- A charged vibration unit, which has rotating vibration in 2 dimensions, produces 2-dimensional or magnetic effects.

- Static electricity has a 1-dimensional charge. Permanent magnets have a 2-dimensional charge.

- Electromagnetism is a 2-dimensional vibration that occurs when a 1-dimensional electron obtains a forward motion.

- Magnetic monopoles do not exist.

- Subatomic particles do not exist because these are vibrations of a lesser complexity, added to an atom.

- Electromagnetism and gravitation are linked together and are a consequence of the number of dimensions in which they exist. In consequence this also applies to the resulting strength.

- Electric and magnetic effects work instantly without an intervening medium, as a result of Space/Time configuration.

2.4 Gravitation

2.4.1 Essence (3-dimensional)

A uniform vector motion does not exert a force. However, in a distributed scalar motion, when a reference frame is applied, the motion is accelerated in the context of the fixed reference frame. This acceleration as a result of distributed scalar motion shows as gravitation on a 3-dimensional object, on matter. Gravitation is scalar and has as such no specific direction.
Gravitation arises as a rotational vibration on a linear vibration on the Progression. The linear vibration has a velocity that is greater than light speed in order to go against the Progression. The rotational vibration is at a speed less than light speed. The effect is gravitation.
This rotational vibration on the linear vibration in 3 dimensions forms atoms, as already explained earlier. Gravitation is thus an inherent property of matter, not a "force"!
Gravitation arises because it is a motion contrary to the Progression, otherwise it is of course no unequivocal motion. Gravitation is thus a motion that is Time-oriented.

2.4.2 Effects

The force aspect of a vector motion is a vector; that of a distributed scalar motion is a field.
Matter does not attract, but all matter goes spatially scalar in the direction of Time. This gives the effect of gravitation.
The more matter (mass) the more gravitation. Gravitation therefore is a local phenomenon, and the effect of gravitation is thus only present where matter is. On the Time side (in the Cosmic sector) we see this as energy and the effect of gravitation here works outward.
So, the more matter there is, the denser it becomes and the less Space units it will occupy. As such more Time units will be released, as less Space means more Time. This then increases the amount of energy available. We will encounter that later also in the explanation of 'why stars radiate'.

As a result of this inward motion we also see that the greater the gravitation, the more Time units there are, and thus the slower clock time goes.

Also, the faster you travel in Space, the slower your clock time runs until you reach the speed of light. This has been proven by American scientists who have flown for 2-weeks in an aeroplane around the world with atomic clocks (accurate to 1 second per billion years!).

This has strange and sometimes opposite effects, for example for satellites that rotate around the earth.

A satellite travels around the globe, so it needs to travel faster the higher it is in orbit above the Earth. The higher it travels the slower the time goes on board compared to the clock on Earth, as a result of its speed. (The famous quote that time runs slower on a mountaintop).

However, on Earth, you are closer to the centre of the Earth. Hence, time runs slower than the clock in the satellite that is in orbit around the Earth, as a result of greater gravitation. In other words, the clock in the satellite goes faster, farther the satellite is removed from the Earth, due to lesser effect of gravitation.

Those two effects, gravitation and speed, work together but do not compensate each other. Recognition of these phenomena is important for the lasting accuracy of our GPS systems, for example.

From this it is clear that clock-time in the Universe is everywhere different, depending on speed and gravitation. On Jupiter for example clock-time runs slower than on Earth. Should an astronaut spend time near Jupiter and later return to Earth he would be younger than if he had stayed on Earth. An effect that has in practice a magnitude of seconds only, unless speeds in the order of light speed or gravitation in the order of black hole strength are involved.

Like the Progression, gravitation becomes visible through the motions of matter. Gravitation therefore is visible as a scalar motion of the atom, inwards to all other Space/Time locations than the one it occupies. That is what all other atoms also do. So it seems like they are attracting each other, which is not the case. Therefore, two masses do not attract each other and Gravitation as force does not exist.

The gravitational vibration of each mass is a scalar vibration on the Progression but greater than the light speed and therefore brings the mass inward in Space/Time.

In other words, the scalar Progression of the Universe is present everywhere, but a material object in it is local and locally creates a change in overall vibration. And this by the scalar character according to the quadratic law.

As long as the gravitation is greater than the Progression, matter goes inwards towards all other Space locations and thus towards each other. That is the case locally in a galaxy.

If the net balance between Gravitation and Progression becomes less, there will be a point that we will call the 'gravitational limit'. When the Progression is greater, all matter continues to move away from each other until they eventually go away with the speed of light of the Progression.

In order to make this clear, imagine a conveyor belt that runs away from a central point. That is the Progression of the vacuum of the Cosmos with the light speed. Cubes and balls are on the belt. The cubes are the photons, which cannot move and go with the speed of the belt. The balls on the other hand can roll. If they roll hard enough, they go against the speed of the belt what demonstrates the Gravitation, but otherwise they go slower than the belt and then move away with the belt.

Gravitation is the controlling factor. If masses go slower than the Progression there is no balance and the Space between the masses (galaxies) grows only bigger and thus the galaxies distance themselves from each other.

In our region of the Universe, where matter through the inward-facing Gravitation moves faster than the outward-facing Progression, that ultimately leads to equilibrium. That balance gives the cohesion to matter.

As Gravitation is an inherent property of matter Gravitation is only manifest where matter is, that is within the gravitational limit.

Galaxies are thus islands in an ocean of Space, which are created as a result of scalar rotational vibrations greater than the speed of light on the outgoing Progression.
This solves the geometry problem that science faces. According to science the gravitational attraction should have resulted in a dense centre of the Universe and gradual reduction in mass the farther out one goes. But this is clearly not the case as the Universe is seen to be rather homogeneous.

It is important to realize that the scalar rotational vibration of an atom moves this atom inward to all locations in Space/Time. And this contrary to the Progression of the vacuum, which moves the location in which the atom is currently located outward in Space. We do not see Space locations, but only the objects that are in it, so we can see that locally all objects move towards each other. For example, we see a mass A in Space moving towards mass B, but in reality, mass A moves towards all Space/Time locations of mass B. Thus, as a mass of 1 Space unit (mass A) moves into Space/Time to a mass of N Space/Time units (mass B) then the force with which they attract each other increases proportionally. This because mass B not only occupies 1 Space location, but also N Time units.
What we see for mass B as motion to 1 Space unit is a move to N Space/Time units. That is the scalar effect of a point for A and a surface for B at a certain radius and thus the quadratic effect for the 'attraction force'.

Due to the scalar effect, the effect of Gravitation is present everywhere and can thus not be shielded. It makes therefore no sense to investigate the possibility of shielding this effect.
As clarified before, a photon has no mass and thus no attraction to another solid object. It's the cube on the conveyor belt of the Progression.

In the *Einstein's Cross*, light from a far-removed galaxy is apparently 'attracted' by another galaxy. Science now takes the view that the galaxy acts as a gravitational lens. And therefore science predicts that a photon must also have a (very small) mass.

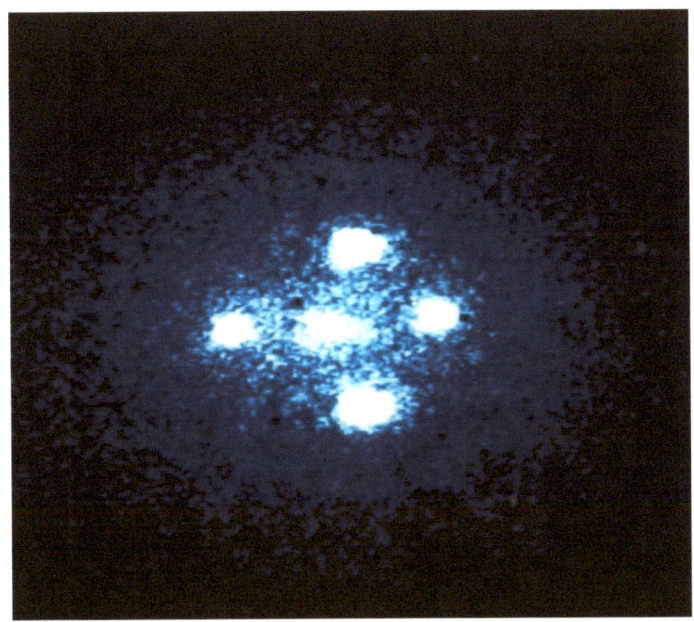

Einstein Cross

However, that is incorrect. The effect finds an explanation in the fact that the gravitational motion of the galaxy is a distributed scalar motion. That is in essence just a decrease in the distance between the two objects. Relatively speaking, it appears as a move from one to another. The total motion of the galaxy is distributed over so much mass that the effect is not visible. The effect of one particle photon however shows. Thus, the direction of gravitation is inward while the direction of radiation is outward. There is therefore no 'gravitational lens'.
The gravitational vibrations are vibrations of the atom with respect to the fabric of Space/Time and no action from one to the other. Gravitation works therefore directly and does not require a medium to transfer the effect.
Even when the distance between masses is astronomical, gravitation works immediately. If a particle would be needed to transfer the 'force' then a time difference would occur and that is not the case.

Therefore gravitation is not transmitted by an elementary particle, a graviton, as science believes. According to science gravitons are tiny, massless particles that mediates the force of gravity. Each graviton would exert a pull on the matter in the Universe, but the particles would be difficult to detect because they interact weakly with matter.

Gravitons and accordingly gravitational waves do not exist in our theory!

This means therefore that the very weak vibrations recorded by the major LIGO-Gravitation wave-meters in the US, in December 2015, in 2017, in April 2019, and in January 2020, in their miles long laser detectors in Washington and Louisiana, must now find another explanation, as was also already the case with previous claims of detected gravitational waves.

Nevertheless, science believes that gravitational waves are caused by the collision of massive objects, such as two black holes or two neutron stars. Maybe these are shockwaves?

2.4.3 Summary and Conclusions

In summary, and in deviation from current scientific theory, we conclude that:

- Gravitation arises as a rotational vibration on a linear vibration on the Progression.

- This rotational vibration on the linear vibration in 3 dimensions forms atoms. Gravitation is as such an inherent property of matter with a Time oriented vibrating mode, hence inward.

- Gravitation then becomes visible as a scalar motion of the atom inwards to all other Space/Time locations than the one it occupies. That is what all other atoms also do, so it seems like atoms are attracting each other. The effect is gravitation. Hence, gravitation as a force does not exist.

- Clock-time (one-dimensional) in the Universe is influenced by speed and gravitation. Hence, clock-time is everywhere different in the Universe.

- As the effect of Gravitation is scalar, Gravitation cannot be shielded.

- Gravitation works without the intervention of a medium such as a curved space and there is no transmission of particles between different masses.

- Gravitation radiations, gravitation waves and gravitons do not exist.

2.5 Atoms

2.5.1 Essence

An atom is the smallest building block of a chemical element. Science calls this a fermion, named after the Italian physicist *Enrico Fermi*, and finds that it is a particle that forms the basis for matter and is characterized by a half-integer spin.
According to science, the atom consists of 3 quarks, 1 so-called up-quark and 2 down-quarks, and the aptly named particles called gluons that bind them together. These then form a so-called hadron of which protons and neutrons are the most important. However, so far science has not been able to prove this model and quarks have not been demonstrated to exist.

Atomic structure (neutron)

Since protons are positively charged particles, a combination of protons in an atomic nucleus would not be possible because they would repel each other. There is therefore a contradiction in the atomic model of science.
This does not occur in our model, because as we have seen an atom is a linear oscillatory vibration in 3 dimensions. Addition of

protons to protons are simply additions of vibration to an existing vibration.
As we have also seen, an atom is formed by adding a 1-dimensional oscillatory vibration to a 2-dimensional rotational vibration. This explains the 2 up-quarks and 1 down-quark.
The oscillating character of the 1-dimensional vibration explains the half-integer spin.
In chapter 2.6 (the chapter on elements) we will describe the exact vibrating build-up for each atomic element.
And as for the gluons, a clarification is given further below.

These vibrations first of all form the proton. According to science, that is the nucleus of a hydrogen atom, around which turn electrons (*Rutherford*).
However, in our model an atom is an integral unit and the special properties of a particular atom are due to the nature and size of the different vibration possibilities that form each atom.
See chapter 2.6 where for each element the vibrational build-up is detailed..

The vibration mode can be explained as follows. A line cannot rotate around itself on an axis because that basically would not constitute a vibration. The linear oscillation thus rotates around its centre. Seen as scalar this vibration delivers in one dimension a vibration like a 2-dimensional disc.
With an additional rotation in a second dimension, it results in a 3-dimensional figure, a sphere. The essence of an atom is thus actually 2 dimensional, to which additional rotations are added to form the elements.

The clarification for the gluons in our model goes as follows.
If two atoms move together under the influence of Gravitation, this can happen only until there is 1 unit of Space left. To get even closer together, a vibration is needed in Time. This vibration (Gravitation) works in Space inwards and in Time outwards. This provides balance because these vibrations work jointly and opposed.

Thus, the net direction of the 3-dimensional rotational vibration of an atom is Time oriented. An atom moving from Space towards Time (Gravitation) moves inward. As soon as the atom enters the Time zone the operation reverses and the atom returns to where it came from. Simply put it collides with itself.
That gives a balance, the atoms go as far as they can.
The same applies to protons, neutrons etc.
As such we have explained the 'binding force' in the atomic nucleus, the strong nuclear force for which science needs gluons. Note that what the gluons from science do not explain is the repulsive effect when a component comes too close to the other component!

From this clarification it follows therefore that there is a clear boundary, namely the unity of Space. To get towards one unit of Space we see a gravitational attraction; when this point is reached repulsion immediately occurs. There is therefore no gradual transition.
When a quantum particle enters the Time zone in the above manner, the reference frame changes.
From "*the less Space, the more Time*" to "*the less Time, the more Space*".
The result will be that for a unit of Time the number of units of Space will be infinite. So for that specific unit of Time, the particle (vibrating mode) will actually be everywhere! Like a message transported on a carrier wave in our daily use of mobile telephones.
That explains the non-local behaviour! So viewed from the Material sector it is a particle, but viewed from the Cosmic sector it is a 'field'. Interference determines the probabilities of spatial location.

An example in this respect can be found in Helium, which first becomes liquid at 269 degrees C below zero. At even lower temperatures, close to the absolute zero it loses its binding to the law of gravitation and can than 'climb' from a cup in which it was sitting. Simply put, it begins to show the reversed aspects of the Cosmic sector where it almost is.

Therefore, atoms are not composed of all kinds of subatomic particles and there are no 'forces' that hold these particles together within an atom. An atom is an integral unit and the special properties of a particular atom are not the result of the components from which it is built but the result of the nature and size of the different vibrating possibilities from which each atom is built.

An atom therefore does not have a 'nucleus' and electrons do not turn around a nucleus. The atom is a system of vibrations. Subatomic particles, such as for example electrons and neutrons, are not atomic components but complex vibrations of the same character as an atom but of a lower degree of complexity. Science, however, sees an atom as composed of parts and names photons gamma particles. And alpha particles are then charged helium atoms and beta particles are electrons.

From the above, that atoms are in fact a collection of interfered vibrations, it follows therefore that changing the composition of materials is not only possible by splitting or fusion of atoms, but also by affecting the vibrating components, for example through interferometry (specific frequency resonance). This latter method is maybe used in the destruction of the twin towers of the NY World Trade Centre on 9/11-2001 (*Dr Judy Wood*) whereby the towers in mere seconds were destroyed into dust.

In the Standard Model of science, there are two types of particles:
- Bosons, which carry force and include gluons and gravitons. These have integer spin (in the theory of this book rotational motion).
- Fermions, which make up matter and include quarks, electrons and neutrinos. These have half-integer spin (in the theory of this book 1-dimensional oscillatory motion).

This Standard Model of modern physics is theoretically probably a good model, but does not adequately represent reality.
In addition, it is incomprehensible and confusing with its bosons, fermions, muons, mesons, baryons, hadrons, gluons, quarks,

antiquarks and strangelets, which each also have attributes such as charge, mass, spin and the like. And to make it even more confusing they have properties that have names like flavour, and colour, charm and beauty, up and down, and strange.
That cannot be elementary!
And in the theory of supersymmetry, there could be many more particles out there awaiting discovery. The theory holds that every particle discovered so far has a hidden counterpart, where each fermion would be paired with a boson, and vice versa.
Science continues to search for the most elementary particle, and many are found. At CERN's Large Hadron Collider (LHC) a 27 kilometres ring buried underground near the border between France and Switzerland, one does their best to find even more of these elementary particles, such as:

- Recently the Higgs boson, with even a Nobel prize attached.
- Or the so-called pentaquark.
- Or an elusive particle, called the chameleon particle, which would have a variable mass, and could apparently help explain both 'Dark matter' and 'Dark energy'.
- Or recently a so-called dark photon that, according to science, indicates a 5th basic natural force!
- Or even small black holes of a mere 100 protons, which would quickly evaporate in so called *Hawking* radiation, developed as an idea by the British cosmologist *Stephen Hawking*. See also chapter 3.4.3 on Quasars.

And on it goes, but none are found that can be considered 'elementary'.
Essentially, in our view one sees only an indeterminate lot of vibrations. Vibrations that occur when random energy particles (vibration modes) are collided with each other. But one can definitely not talk of elementary particles.

2.5.2 String Theory

When physicists (*Edward Witten*) assume all the elementary particles are actually one-dimensional loops, or "strings," each of which vibrates at a different frequency, science gets much easier apparently. String theory allows physicists to reconcile the laws governing particles, called quantum mechanics, with the laws governing space-time, called general relativity, and to unify the four fundamental forces of nature into a single framework.

But the problem is, string theory can only work in a Universe with 10 or 11 dimensions: three large spatial ones, six or seven compacted spatial ones, and a time dimension. The compacted spatial dimensions — as well as the vibrating strings themselves — are about a billionth of a trillionth of the size of an atomic nucleus. There's no conceivable way to detect anything that small, and so there's no known way to experimentally validate or invalidate string theory.

To put the strangeness of string theory further in perspective it developed into M theory. M theory views the universe as a 3-dimensional bubble floating in a multi-dimensional world of trillions of other bubbles, each with its own events etc. This theory apparently can be proven when 'Dark matter' can be proven to exist. Success!

In the theory of this book string theory may be a nice mathematical modelling exercise, but cannot be a serious cosmological model if it needs 10, 11 or even 26 dimensions and cannot be validated. And the essential difference between string theory and the theory of this book is that science needs something (strings) to vibrate whereas for the theory of this book vibrations as motion in itself is fundamental, vibrations of S/T.

String theory was initially developed to explain the strong nuclear force, for which gluons were thought to be the binding particle. This after cyclotrons found that protons and neutrons could be broken down and many more so called fundamental particles were found.

However, in the theory of this book the strong nuclear force has a simpler explanation, as clarified earlier. To state again:
If two atoms move together under the influence of Gravitation, this can happen only until there is only 1 unit of Space left. To get even closer together, a vibration is needed in Time. This vibration (Gravitation) works in Space inwards and in Time outwards. This provides balance because these vibrations work jointly and opposed.
Thus, the net direction of the 3-dimensional rotational vibration of an atom is Time oriented. An atom moving from Space towards Time (Gravitation) moves inward. As soon as the atom enters the Time zone the operation reverses and the atom returns to where it came from. Simply put it collides with itself.
That gives a balance, the atoms go as far as they can.
The same applies to protons, neutrons etc. As such we have explained the 'binding force' in the atomic nucleus, the strong nuclear force for which science needs gluons.

Note that there is another problem science faces with their explanation for gluons as a binding force. Because gluons do not explain the repulsive effect that arises when a component comes too close to the other component!

2.5.3 Summary and Conclusions

In summary, and in deviation from current scientific theory, we conclude that:

- The atom is a system of vibrations, formed by adding a 1-dimensional vibration to a 2-dimensional vibration.

- Subatomic particles are not atomic components but complex vibrations of the same character as an atom but of a lower degree of complexity.

- The properties of a particular atom are the result of the nature and size of the vibrating composition of the atom.

- The cohesion in the atomic structure is not dependent on electrically charged components, but results from Space/Time configuration, from the balance achieved in the number of dimensions of the scalar rotation.

- The charge of an atom (ionization) is not dependent on the number of electrons orbiting it, but is dependent on the net atomic rotation, in Space or in Time.

- There are no fundamental forces of nature, but such forces manifest themselves through distributed scalar motions.

- As an atom is a system of vibrations matter can also be changed by interferometry.

- The Standard Model of science is not a correct representation of reality, as it is based on material properties.

- As such the search for elementary particles is open-ended.

- String theory is an incorrect model of reality.

2.6 Chemical Elements

2.6.1 Essence

Molecules are formed when two or more atoms join together. A molecule then forms a substance of matter and is therefore the smallest particle of that substance that still has the properties of that substance. That forms a chemical element.
As previously clarified, these chemical elements are formed by an added rotation to a 2-dimensional rotating unit. That rotation is always directed towards the Cosmic sector, because matter goes in Time (Gravitation).

Added rotational options are, for example, the following.

- The 2-dimensional unit can rotate in its entirety around the third dimensional axis.
- Also, two of these 2-dimensional units can rotate together as a whole around the third axis. In that case, they must both have the same rotational speed to prevent interference. Generally, they then rotate counter-directed.
- Addition of more rotational units to these two 2-dimensional rotating systems increases the rotation of both in 1 dimension, instead of the rotation of 1 system in both dimensions. This to prevent interference.

In those cases where the vibrations in the two dimensions are different, the main rotation is called the principle rotation and the other is the subordinate rotation.
Then there is still the rotation in 1 dimension, the electric rotation. It is added in the opposite scalar direction, in Space or in Time.

Based on this, a system has been developed that allows rotation of atoms to be displayed.
The system gives 3 numbers, the first of which represents the principle 2-dimensional magnetic rotation, the second number the

subordinate 2-dimensional magnetic rotation and the third number the 1-dimensional electric rotation.
The electric rotation is a 1-dimensional vibration directed in Space or in Time. Directed in Space is indicated by placing it in brackets. Please note that charge here is not applicable!
And again a magnetic rotation is thus a 2-dimensional vibration.

Example:
   1-0-(1) = Electron
   1 = The principle 2-dimensional magnetic rotation
   0 = The subordinate 2-dimensional magnetic rotation
   (1)= the 1-dimensional electric rotation, with brackets indicating Space directed ionization, no brackets is Time directed ionization.

This gives the following:
   Electron   1-0-(1)
   Positron   1-0-1
   Neutron   1-1-0
   Neutrino   1-1-(1)

Neutrinos are thus 2 rotating magnetic rotations (a neutron) with 1 added electric rotation in Space.
Neutrinos are produced in very large quantities in the Universe as Cosmic radiation from the Cosmic sector of the Universe. In the uncharged state they go through everything, because they are directed in Space.
For example, 70 billion neutrinos per second pass through every square centimetre on Earth.
However, when neutrinos sometimes obtain a charge when they go through matter they cannot get out any more. Although the chance of that happening is estimated to be 1 to 10 billion, there are a lot of neutrinos and the concentration of charged neutrinos in matter is gradually increasing, which results in ageing of matter. In this way galaxy systems age as well, as we will see later.

Addition of 1 magnetic rotation unit to a neutron results in:
   Atomic No. 2   2-1-0 Helium

Further additions of magnetic rotation units yield the noble gases in succession:

Atomic No. 10	2-2-0 Neon
Atomic No. 18	3-2-0 Argon
Atomic No. 36	3-3-0 Krypton
Atomic No. 54	4-3-0 Xenon
Atomic No. 86	4-4-0 Radon
Atomic No.118	5-4-0 Unstable

These rotations around 3 axes thus produce the elements. These are the 117+ chemical elements of the periodic table. See below.

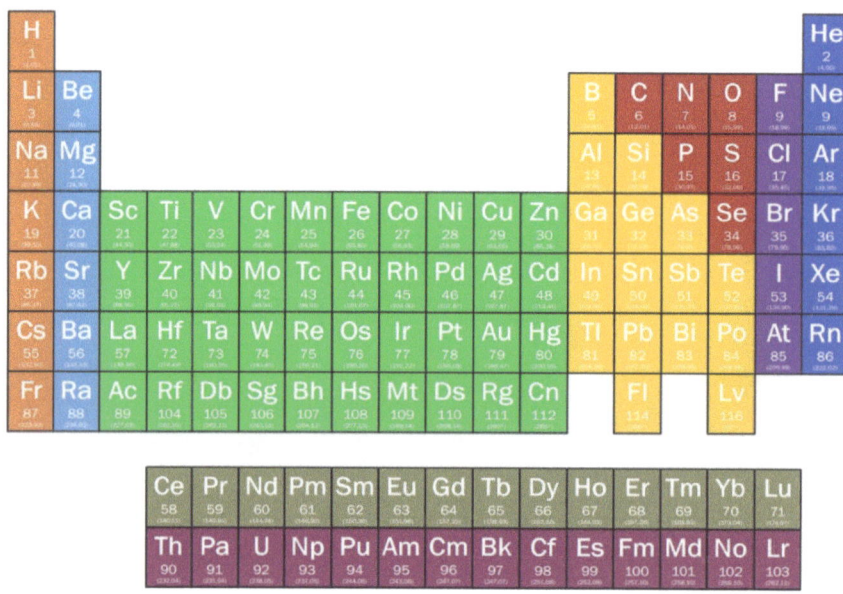

The Periodic System of Elements

In the case of an excessive amount of additions, such a combination of vibrations becomes unstable and falls rapidly apart. This instability results in what we call radioactivity.

The explanation for this is the fact that matter is directed in Time, towards the Cosmic sector, where activities operate reversed. Should an atom become too complex it will explode.

Radioactivity is an explosive phenomenon of matter in the Cosmic sector. These are random explosions of atoms in Time. An explosion in Space begins somewhere and propagates out at high speed. In a radioactive explosion the action starts somewhere in the matter and propagates outside in Time, with a reverse speed (therefore slow), and as such spread over a long time.
This explains what science calls the 'half-life' of radioactive elements.
And it also explains the weak nuclear force (hence, no W and Z bosons, as science thinks).
We will encounter explosive action of matter again on the Cosmic scale.

With multiple magnetic rotations in an atom, a variable number of electric rotations can be added. Addition of multiple magnetic rotations to an atomic nucleus, with the addition of a variable number of electric rotations, thereby creates the periodic system of the elements.

The number of electric rotations is determined by the relation $2 \times N^2$, i.e. 2-8-18-32. The explanation for this is that if we model a nucleus as two 2-dimensional rotations (magnetic) with the same speed (otherwise interference occurs) then the 1-dimensional rotation (electron) of a N number of nuclei can mathematically have the relation $2 \times N^2$. Double because the spin can have 2 directions.
So, matter is always directed towards the Cosmic sector, hence rotates in Time.
Thus, a 2-dimensional magnetic rotation is directed in Time, while an electric rotation is a 1-dimensional rotation in Space or in Time. That means that a magnetic rotation is equivalent to $2 \times N^2$ electric rotations.

An electron is modelled as 1-0-(1). The first real element is Helium 2-1-0. By adding an electric rotation, Hydrogen is produced 2-1-(1) by adding 1 equivalent electric rotation in Space.
A rotation in Space is shown by placing it in brackets, a rotation in Time is shown without brackets.
Thus, in order to build up the elements from Helium, a magnetic rotation is modified with one or more electric rotations.
For N = 2, the equivalent is $2 \times N^2 = 8$ electric rotations for a magnetic rotation and a group of 8 elements is created.

The first 4 elements after Helium thus arise by adding 1 additional electric rotation in Time as follows:

    Atomic No. 2   2-1-0 Helium
    Atomic No. 3   2-1-1 Lithium
    Atomic No. 4   2-1-2 Beryllium
    Atomic No. 5   2-1-3 Boron
    Atomic No. 6   2-1-4 Carbon

The fourth element Carbon can also be formed by the addition of 1 magnetic rotation and 4 electric rotations in Space as follows:
    Atomic No. 6   2-2-(4) Carbon

The following elements are then formed by the addition of 1 magnetic rotation and reduction with electric rotations in Space, as follows:

    Atomic No. 7   2-2-(3) Nitrogen
    Atomic No. 8   2-2-(2) Oxygen
    Atomic No. 9   2-2-(1) Fluor
    Atomic No.10   2-2-0   Neon

Then follows another such group with the addition of 1 magnetic rotation.
    Atomic No.18   3-2-0 Argon
Argon ends this sequence of elements.

As a result, two units are added in both magnetic dimensions, so the following group must be N = 3 and thus $2 \times N^2 = 18$ elements, by adding 18 electric rotations.

Then again a group in which the magnetic rotation is greater by 1. This results in 4 magnetic rotations in one dimension and now there are 32 elements in each of the following 2 groups.

With 5-4-0 there are 4 units above the 1-0-0 date in both magnetic dimensions and the elements now become unstable. We are then at Atomic no. 117.

With increasing magnetic rotations, the energy content (frequency, colour) of the electric rotation must also increase proportionally. That explains an atom's 'spectrum' (*Niels Bohr*), which appears as spectral lines in light.

In this way, an atom may change from one pattern to another by absorbing or emitting certain units of vibration (energy), which is observed as absorption or emission of photons.
Remember: *The photon is the basic carrier of energy.*

A special effect is created by inclusion of neutrinos in the vibration pattern of an atom. As mentioned, neutrinos are particles that can have a magnetic charge.
Remember neutrinos are 2 rotating magnetic rotations (a neutron) with 1 electric rotation in Space. Normally, neutrinos go through matter. But should they accidentally get a rotation in Time, they are absorbed into the atom and will mix into the vibration pattern of the atom.
As such, it contributes to the properties of the atom and in this case to the mass. That produces the isotopes.
Isotopes are therefore an ageing phenomenon of atoms.

2.6.2 Summary and Conclusions

In summary, and in deviation from current scientific theory, we conclude that:

- Chemical elements are formed by an added rotation to a 2-dimensional rotating unit, in direction of Time.

- A system has been developed that allows rotation of atoms to be displayed simply and elegantly that construct elements.

- The periodic system of the elements in the system is formed by adding a variable number of 1-dimensional electric rotations to the multiple 2-dimensional magnetic rotations in an atom.

- Radioactivity is an explosive phenomenon of matter in the Cosmic sector. Such random explosions of atoms in Time propagate in a manner opposite to an explosion in Space.

- In radioactivity the speed is therefore slow and spread over a long time. This explains the 'half-life' of radioactive elements and the weak nuclear force (no W and Z bosons).

- A magnetic rotation is equivalent to $2 \times N^2$ electric rotations.

- Isotopes are an ageing phenomenon of atoms. Neutrinos are produced in the Cosmic sector as cosmic radiation and when neutrinos are absorbed into an atom they will mix into the vibration pattern of the atom and as such form isotopes.

- The spectrum of an atom, the spectral lines in light, show the energy content as a result of the number of magnetic and electric rotations. An atom may change from one pattern to another by absorbing or emitting units of vibration (energy), which is observed as absorption or emission of photons.

2.7 Molecules

2.7.1 Essence

Molecules arise because atoms form one or more connections with each other. Such a connection will be formed when atoms share a part of their electric rotations as standing waves, when they are close enough to each other in Space.
The mass of the molecule is the combined mass of the atoms of which it consists.
In the following, we will globally look at bonding, shape and properties.

2.7.2 Bonding

We already saw that when two atoms join together the atoms can only come as close as 1 unit of Space left. So it follows that there is a clear boundary, namely the unity of Space. To get towards one unit of Space we see a gravitational attraction; when this point is reached repulsion immediately occurs. There is no gradual transition.
As a result the equilibrium in a solid is achieved in all 3 dimensions. The cohesion of atoms within a solid can be influenced by supplying energy (opening the Time tap) with for example thermal energy (heat). Thermal energy (Time) input is directed opposite to the Progression thus increasing vibration amplitude. The atoms thus move further apart resulting in breakage of the vibration creating the bond.
At a certain quantity, the scalar effect of the vibration is eliminated in 1 dimension, the melting point. The atom is still attached to its neighbours in 2 dimensions, but can now move freely in one dimension. It is now liquid.
The situation regarding the evaporation point is comparable.
And in the gas phase, a molecule can move completely free outside one unit of Space.
The difference between the solid, liquid, and gaseous phase is thus dependent on the energy content.

It is a phase change and not a chemical reaction as science advocates, as we will review hereafter.

As a result of the 3-dimensional nature of matter there are thus 3 phase transitions. (Plasma as ionized gas is not a true phase transition in our model because of its 2-dimensional character).
A phase change occurs under normal conditions at a clear and clearly defined energy level for a pure substance. Impurities in the substance can be detected in this way. Volume and pressure can affect the required energy level.
In the theory of this book the bonding between atoms is thus a feature of the atom itself, in its dimensional build-up. These result from standing wave vibrations in 3 scalar dimensions. This clarifies the sudden merging and snapping of the bond. As clarified before there is no gradual transition.
Therefore, when two atoms join together the atoms can only come as close as 1 unit of Space left. To get even closer, another one-dimensional electric vibration is needed, but now in Time. One vibration now works in Space inward and another in Time outward. These give balance because these 2 vibrations work in opposition to each other. This creates a standing wave.

A standing wave is a wave phenomenon caused by interference of two waves of equal frequency and amplitude but in opposite propagating direction. This results in a regular pattern of points that do not move, the nodes, and points that show maximum amplitude, the bellies. The distance between the nodes is the half wavelength of the interfering wave. The frequency determines the energy content of the bonding connection.
Because this 1-dimensional electric vibration as a standing wave bonds the atoms, it gives the effect of a spring. The atoms cannot separate without breaking the harmonic wave. That gives the bonding energy.

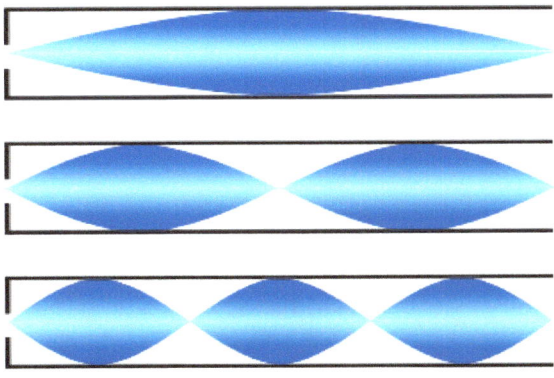

Standing Waves

The upper figure is called the first harmonic, next the second, and so on.
The first harmonic provides the strongest bond.
The strongest bonding energy is found in compounds consisting of the same atoms, which in this way form molecules of an element.
The bonding force is the energy content of the harmonic wave.
Some bonding forces are strong as in salts, others are weaker.

This clarification for cohesion in matter is seen as a better explanation than what science advocates, which is a chemical bonding, based on positive and negative charges.
According to science, in order to be able to bond, atoms must first share electrons and achieve a balance between the positive and negative tug created by their nuclei and electrons. This will increase their stability and lower their energy.
The length of chemical bonds are believed to be between 0,1 and 0,3 nanometre, about a half million times smaller than the breath of a human hair. This is very difficult to perceive. Nevertheless, science have made it possible.
The website www.inverse.com/article/62367-watch-first-video shows a video of two atoms bonding and breaking.
The 18-second video shows 2 atoms bobbing around separately in hollow carbon nanotube cylinders. But slowly they bob closer and come together, before the two tiny specks suddenly merge as one. The unified figure then starts to distort, jumping about at different

angles and the bond snaps when the length of the bond exceeds the size of the atoms.

2.7.3 Form

Such molecular compounds can form crystals in pure solid form, and in patterns that usually represent the underlying pattern of the molecule itself, 2 or 3 dimensional.
That results in 7 crystal systems: cubic, tetragonal, hexagonal, trigonal, orthorhombic, monoclinic, and trichlien. So 7 possibilities of regular shape.
Thus, in a crystal, the molecules are arranged in a solid geometric pattern.
Crystals are unique in that they are good conductors of light, sound and electricity. Piezoelectricity is an effect in crystals whereby an electric charge accumulates as a result of mechanical stress. (Example: lighting a BBQ).
It is a reversible process, meaning that application of an electrical field will result in mechanical strain. Granite for example has peculiar properties as a result of quartz crystals embedded in rock. This effect is apparently also occurring in DNA and certain proteins, reason maybe why crystals are used in certain healing practices.

In a crystal of isotropic elements, all atomic bonding forces are therefore the same. This is in contrast to substances consisting of different elements which then have different bonding forces within the same substance.

As an example of a crystal structure we take Carbon, of which there are 2 forms of crystalline compounds, namely diamond and graphene.
For the 3-figure notification indicated next to the crystalline compound the system for rotational vibration structure as explained in chapter 2.6 is used.

- **Diamond 2-2-(4)**
  Hereby the carbon atoms are arranged in a pattern where the core atom is at the centre and has 4 standing waves that bond with 4 other carbon atoms. In addition, the atoms are arranged as a 3-dimensional tetrahedron. This gives a very strong bond, making diamond so hard.

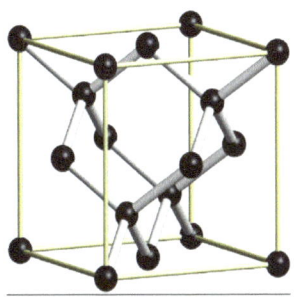

Diamond structure

Therefore, diamond is also transparent because photons can not supply enough energy to change the pattern. If there are other elements included as impurities, diffraction occurs and with it discoloration and the diamond is thus considered impure.

There are also black diamonds, the so-called Carbonados. These are extremely rare, do not contain earthly inclusions and are more porous than normal diamonds. It is suspected that they originate from supernova remains, and they are thus of extraterrestrial origin.

Although diamonds on Earth are rare, extraterrestrial diamonds are very common. It is believed that in the atmospheres of Uranus and Neptune diamonds rain, while the asteroid Steins, visited by NASA in 2008 (Rosetta), is shaped like a 5 km diameter diamond.

- **Graphene 2-1-4**
  Hereby the carbon atoms are arranged in a pattern where the core atom is at the centre of 3 standing waves that bond

with 3 other carbon atoms. In addition, the atoms are arranged in a 2-dimensional 6-sided honeycomb pattern.
It is a very tight bond but only in the flat plane. This produces a very tough and strong material.
For example, a layer of one atom thickness is strong and flexible enough to serve as a hammock for the weight of a car.

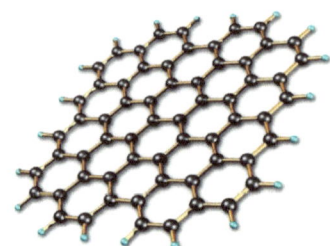

Graphene structure

Based on graphene, among other things, the following materials are formed.

- **Graphite**
  Graphene in layers together form graphite. The layers are not tightly bound and can thus be easily dislodged. As a result, it can be used as a lubricant or as a pencil.
  It is also used as a neutron moderator in nuclear reactors, as it is thermally stable and conducts heat very well. .

- **Bucky balls**
  These are formed when a sheet of graphene is folded into a ball, which is possible with a large number of atoms (for example 60 or more). Bucky balls can be created by lightning strikes.
  Bucky balls are named after *Buckminster Fuller*, an architect who designed buildings that had such a form.

Bucky ball structure

- **Nanotubes**.
  These are formed when a sheet of graphene is folded into a tube.

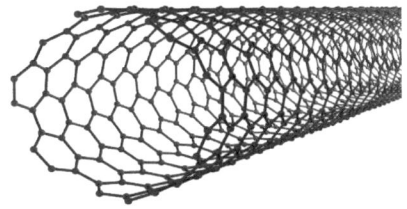

Nanotube structure

When carbon atoms are interconnected as a chain with branches, we call them isomers.

In combination with other elements organic compounds are formed, such as the hydrocarbons (petroleum products), but also, for example, alcohols, esters, ethers, sugars, and the like.

Compounds with different elements are composite molecules and form substances with different bonding (energy content) such as water, which consists of hydrogen and oxygen.

Due to the different elements in it, multiple crystal lattices are possible. For water think of the different shapes in ice crystals and snow.

Compounds are possible in 1, 2 or 3 dimensions and as such form a gas, a liquid or a solid. These we call phase changes.

Substances need not necessarily have a chemical bond but may also be formed in a mixture. After condensation and solidification the different atoms also rank here in a grid.

2.7.4 Properties

This provides the ability to influence the properties of a substance, for example by rapid cooling and thereby maintaining the resulting grid. In example: Quenching and tempering treatments such as for forgings.
Phase diagrams have been developed for this for different materials.

On the basis of composition, bonding strength, shape and the like, an infinite number of materials with different properties become possible. That results in natural materials but also artificial materials can be made in this way, such as polymers.

2.7.5 Summary and Conclusions

In summary, and in deviation from current scientific theory, we conclude that:

- Molecules are formed when atoms share a part of their electric rotations as standing waves as a connection between them.

- Bonding energy results from a 1-dimensional electric vibration as a standing wave with the effect of a spring. The atoms cannot separate without breaking the harmonic wave.

- As a result of the 3-dimensional nature of matter there are 3 phase transitions, solid (3-dimensional bond), liquid (2-dimensional bond) and gas (1-dimensional bond).

- Application of thermal energy will increase vibration amplitude and as such weaken the dimensional bonds, one by one over the 3 dimensions. This is called phase transition.

- Phase transition thus results from elimination of the scalar effect of rotation in 1 or more dimensions, and is dependent on energy content. It is not a chemical bonding, based on positive and negative charges, as advocated by science.

Chapter 3   Cosmology

The Universe wants to make itself known to those who can comprehend its language, and that language becomes more and more intelligible to us as our spiritual component unfolds
Yitzhak Bentov

3.1 The Universe

3.1.1 General

In the theory of this book the Universe consists of 2 sectors, the Material sector and the Cosmic sector, whose property is scalar motion, as explained in Chapter 1 on Physics.
Scalar means motion in all directions, and shows different results in measurement of properties, when subject to translation motion in a reference system.
For science there is only the cosmos, which in the theory of this book is seen as the Material sector of the Universe.

When we look up at the sky on a clear night we see a multitude of stars. All these stars without exception belong to our galaxy, the galaxy that our Sun is also part of. Just 100 years ago, the cosmos consisted only of our galaxy, the Milkyway. Since then, around 500 billion galaxies have been found and it is still ongoing.
The realization that we live in a galaxy which is one among many galaxies came only with the use of large telescopes, as the light of these galaxies is not visible to the naked eye. And with this realization came the recognition that the cosmos is enormous.
To put this in perspective the following. Our galaxy has a diameter of roughly 100,000 lightyears, meaning that it will take 100,000 years for light to travel from one end of the galaxy to the other end, at the speed of light (200,000 km/sec). If we represent this diameter by a 10 mm diameter piece, such as a 10 cent coin, we see the next galaxy (the Andromeda galaxy) as another 10 cent coin at a distance of 20 cm. In this representation galaxies are distributed at such distances in all directions, with 1 meter gaps in

between here and there in our representation. The most distant galaxies will then be located at 1300 meter distance, representing 13 billion lightyears.
As a side issue, note that travelling between galaxies would be difficult from a navigation point of view, as there are no fixed points to navigate on, no stars, no light.

Measurements of the cosmos by science are done on matter. And in doing so science sets a frame of reference. Science then finds that galaxies that are nearer to us apparently move away from us slower than galaxies that are farther away from us. And the galaxies farthest away from us seem to move away at the speed of light.
On basis of distance measurements in combination with redshift measurements of frequency of light, science assumes that the cosmos shows acceleration and thus expands. As such science have concluded that there must be 'Dark Energy' to produce the force for such acceleration.

Where our theory can explain the expansion of the cosmos from the scalar nature, in our view a problem arises for science in their measurements of speed on basis of redshift measurements. We will further clarify this now by reviewing how science measures distance and speed.
Distance of stars in our own galaxy can be measured by triangular measurements because these are relatively close. In addition a method for measuring distances of stars has been determined on basis of known luminosity of certain stars, the socalled cepheïds. This is all merely straightforward. But distance of a cosmological object is more difficult. Such distances are measured by measuring light of a type 1 novae, which also have a known luminosity and as such, like cepheïds, can be used as a standard candle. Knowing the absolute magnitude, the distance can then be calculated.

Measuring speed of a cosmological object is done differently. Speed of a cosmological object that is at right angles with the viewing direction is measured by determining the difference in

position relative to other objects. Speed of an object in the viewing direction is measured cosmologically with the Doppler-effect, or in other words the redshift (or blue shift) in its spectrum of light.
The Doppler-effect is the frequency change that occurs depending on whether or not the object moves towards (blue shift) or away (redshift) from the observer. (Similar to the sound of a car siren which changes with the direction of observation).
These measurements show that galaxies are apparently moving generally away from each other. Due to this finding science concludes that the cosmos expands. For galaxies moving away from us the change in frequency of their light must be to the red part of the frequency band
When science next combines distance with speed, on basis of redshift measurement, a problem arises. The farther away a galaxy is, the faster the galaxy apparently moves, while the galaxies that are farthest away seem to move away at light speed. Galaxies thus appear to accelerate, leading to the conclusion that the Universe expands.
For this explanation by science that the Universe expands, a Nobel prize was awarded in 2011 (to *Perlmutter, Brian Smith, Adam Riess*) with the interpretation that the Universe accelerates, and therefore there must exist 'Dark Energy' to explain this acceleration.

In the theory of this book the above conclusion is incorrect. The apparent acceleration is the effect of distributed scalar motion as explained in Chapter 1 on Physics.
The Progression moves in scalar mode at light speed all galaxies away from each other. The Gravitation moves in scalar mode in general all galaxies in opposite direction. The farther a galaxy is away from us the less the relevant visual effect of the gravitation is to us. That has as result that the galaxies close to us move at speeds less than light speed and those farthest away from us move away at the speed of the Progression, thus at light speed. There is therefore no acceleration, all galaxies move at light speed with the Progression and have an opposite directed speed from Gravitation, depending upon their mass. The farther a galaxy is

away from us the less the effect of Gravitation becomes visible to us.

This is therefore a different explanation than the explanation from science and it does not require "Dark Energy" as science must introduce to explain the apparent expansion of the Universe. Moreover, there is a problem in the manner science uses the redshift measurements to determine speed as will be clarified in the next chapter 3.1.2 on Dark Energy.

3.1.2 Dark Energy

The accelerated expansion was discovered during 1998, by two independent projects, the Supernova Cosmology Project and the High-Z Supernova Search Team, which both used distant type 1a supernovae to measure the acceleration. The idea was that all type 1a supernovae have almost the same intrinsic brightness (a standard candle). And since objects that are further away appear dimmer, we can use the observed brightness of these supernovae to measure the distance to them. The distance can then be compared to the supernovae cosmological redshift, which measures how much the universe has expanded since the supernova occurred.

The expansion of space came with the realization by *Hubble* that there is a relation between velocity and redshift. To quote Wikipedia:

Quote

> *By comparing the apparent brightness of distant standard candles to the redshift of their host galaxies, the expansion rate of the universe has been measured to be $H_0 = 73.24 \pm 1.74$ (km/s)/Mpc. This means that for every million parsecs of distance from the observer, the light received from that distance is cosmologically redshifted by about 73 kilometres per second (160,000 mph). On the other hand, by assuming a cosmological model, e.g. Lambda-CDM model, one can infer the Hubble constant from the size of the largest fluctuations seen in*

> the Cosmic Microwave Background. A higher Hubble constant would imply a smaller characteristic size of CMB fluctuations, and vice versa. The Planck collaboration measure the expansion rate this way and determine H_0 = 67.4 ± 0.5 (km/s)/Mpc. There is a disagreement between the two measurements, the distance ladder being model-independent and the CMB measurement depending on the fitted model, **which hints at new science beyond our standard cosmological models** (my highlight).

Unquote

As an aside comment it should be noted that a constant of 67.74 km/s per Mpc would lead to an age of the Universe of 13.8 billion years, whereas one of 73 would indicate a Universe age no greater than 12.7 billion years. It is a mismatch that suggests that there are stars (such as HD 140283) that are apparently older than the universe (?). Again, a hurdle for science to overcome in their theory.

In the theory of this book the cosmological redshift comparison does not account for speeds nearing and exceeding light speed. For example, part of the explosion product of a nova, being used as standard candle, exceeds light speed, as will be explained in chapter 3.4.1 on novae.
This also applies to quasars, who in the view of science are extremely far away and extremely bright on basis of the redshift. In my theory however novae and quasars do exceed light speed and as such can show redshifts over 1 while not necessarily being extremely far away.
As will be seen later, at speeds greater than the light speed, part of the explosion energy is used to overcome gravitation and the other part of the speed goes in Time and thus opposite Space. That means that the faster a nova or quasar goes, the less distance it covers. The more Time the less Space.
Hence, in the interpretation of this theory here, there is no acceleration of the galaxies in the Universe but there is a constant movement, called the Progression, with the speed of light. There is no 'Dark energy' that would be necessary to explain an apparent

acceleration and it is therefore not surprising that science cannot find 'Dark energy'.
Science will therefore have to replace the accelerating reference framework with distributed scalar motions, of Progression outward and Gravitation inward.
The problem for science starts with the assumption that the universe expands. The expansion then needs a force to do so, the *"invisible agent that pushes space-time apart"*. 'Dark energy' is therefore an example of an "auxiliary hypothesis", an ad hoc postulate that is added to a theory in response to observations that falsify it.

3.1.3 Big Bang

Hereunder a picture of the Big bang theory.

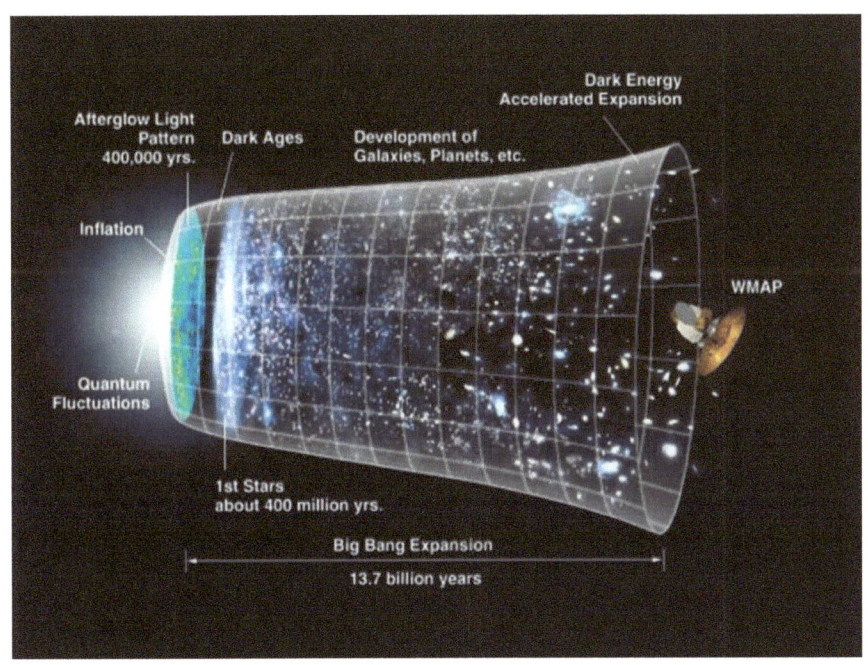

The (incorrect) Big Bang theory pictured

As the expansion of the Universe in our theory is scalar, interpolation to a point far in the past where all galaxies were

apparently at one point together, as science advocates and thereby propagating the origin of the Universe to be the outcome of a Big Bang, is seen as incorrect.
To overcome a number of problems with the Big Bang theory an additional inflation theory was developed.
These problems included in example a flatness problem (the geometry of the Universe at large scale appears Euclidean) and a horizon problem (the Universe appears statistically homogeneous and isotropic).
For this inflation theory a Nobel prize was issued (*Alan Guth*).

To put the inflation in perspective a quote from Wikipedia.

Quote
>During the inflationary epoch about 10^{-32} of a second after the Big Bang, the universe suddenly expanded, and its volume increased by a factor of at least 10^{78} (an expansion of distance by a factor of at least 10^{26} in each of the three dimensions), **equivalent to expanding an object 1 nanometre (10^{-9} m, about half the width of a molecule of DNA) in length to one approximately 10.6 light years** (about 10^{17} m or 62 trillion miles) long. A much slower and gradual expansion of space continued after this, until at around 9.8 billion years after the Big Bang (4 billion years ago) it began to gradually expand more quickly, and is still doing so.

Unquote

Such inflation clearly exceeds all laws of science, and becomes unbelievable and a problem in itself!

In the theory of this book the Universe consists of scalar motions of Progression and Gravitation, the effects of which become apparent in the motion of matter (refer Chapter 1 on Physics).
A Big Bang theory (*Lemaitre*), coupled with a necessary but unbelievable inflation theory, is therefore not necessary.
As we will see later, matter from the Material sector disappears to the Cosmic sector. The Cosmic sector eventually returns that matter to the Material sector, which we then see appearing in the

Material sector as cosmic radiation. This leads to a cyclic Universe.

3.1.4 Cosmic Radiation

Cosmic radiation consists of particles (units of vibration) that apparently do not come from a particular source in the Material sector but appear to come from all sides, spread out at speeds near light speed. Cosmic radiation originates in the Cosmic sector. Upon appearance in the Material sector these particles are quickly converted into other types of particles in a very short time. These are stable particles in the Cosmic sector but very unstable in the Material sector. It is the Cosmic Microwave Background (CMB) radiation of 2.7 degrees Kelvin, whose existence was discovered first in 1964 (*Penzias and Wilson*).

The image below gives an image of this radiation in the microwave frequency and is colour coded, with red the highest intensity and blue the lowest. Subsequently, the general average has been eliminated and the whole is multiplied by 10,000. The different colour patterns thus give minimal variations of the mean!

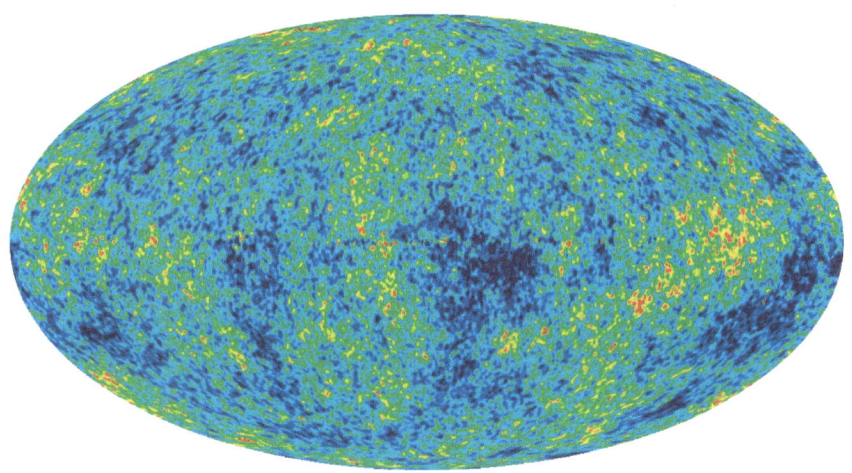

Cosmic radiation

This shows that the Cosmic radiation is extremely isotropic. That means that it is distributed equally and evenly throughout. Science sees this radiation as evidence of the 'Big Bang', assuming that the cosmos was created by an explosion of a singularity, a point of enormous density.

A problem for science here, however, is that this radiation is so isotropic, distributed evenly throughout space. That makes it difficult to explain how galaxy systems now appear in clusters and superclusters and are not distributed homogeneously.
As stated before, to find an explanation for this discrepancy, science has come up with the crazy idea of hyperinflation in addition to the Big Bang, in which the Cosmos increased in a very short time in volume and at speeds greater than light speed.
The theory of rapid inflation was thus introduced to explain the reduction of the amount of mass per unit volume.
But such incredible scientific theory is unnecessary in the interpretation provided in this book.

The existence of a mirror image Cosmos (the Cosmic sector) also explains why there is only matter in Space and hardly any anti-matter. That is also a problem for the fans of the Big Bang theory because, according to these supporters, both forms of matter should have been formed to the same extent in the Big Bang.
According to science the question of why there is so much more matter than its oppositely-charged and oppositely-spinning twin, antimatter, is actually a question of why anything exists at all.
Science assumes the Universe would treat matter and antimatter symmetrically, and thus that, at the moment of the Big Bang, equal amounts of matter and antimatter should have been produced.
But should that have happened, there would have been an instantaneous and almost total annihilation of both.
Protons would have cancelled with antiprotons, electrons with anti-electrons (positrons), neutrons with antineutrons, and so on, leaving behind a dull sea of photons in a space without matter.
For some reason, there was excess matter that didn't get annihilated!

The explanation is that anti-matter exists in the Cosmic sector and is distributed in Time and not in Space. It is therefore not present in Space in concentrated form but spread out.

Therefore, if in a cyclotron anti-matter is generated in our Material sector, it can only be produced in very small quantities and the result is extremely volatile. There is thus no danger that creation of anti-matter in a cyclotron could destroy our Earth, as some fear.

3.1.5 The Nature of the Universe

The problems that science encounter to explain the nature of the Universe can be demonstrated with the following.

The nature of the Universe strongly depends on a factor of unknown value: Ω, a measure of the density of matter and energy throughout the cosmos.

- If Ω is greater than 1, then space-time would be 'closed' like the surface of an enormous sphere. If there is no 'Dark Energy', such a Universe would eventually stop expanding and would instead start contracting, eventually collapsing in on itself in an event dubbed the 'Big Crunch'. If the Universe is closed but there *is* 'Dark Energy', the spherical Universe would expand forever.

- Alternatively, if Ω is less than 1, then the geometry of space would be 'open' like the surface of a saddle. In this case, its ultimate fate is the 'Big Freeze' followed by the 'Big Rip'. First, the outward acceleration of the Universe would tear galaxies and stars apart, leaving all matter frigid and alone. Next, the acceleration would grow so strong that it would overwhelm the effects of the forces that hold atoms together, and everything would be wrenched apart.

- If $\Omega = 1$, the Universe would be flat, extending like an infinite plane in all directions. If there is no 'Dark Energy', such a planar Universe would expand forever but at a continually decelerating rate, approaching a standstill. If there is 'Dark

Energy', the flat Universe ultimately would experience runaway expansion leading to the Big Rip. Regardless how it plays out, the Universe is dying.

Clearly, scientists still wonder why the Universe is so precisely adjusted that certain physical constants in the laws of nature, such as Ω, have values that by the slightest deviation would not have created a Universe, or a totally different Universe.
These constants all disappear in the theory of this book, as we will further clarify in chapter 3.5.2.
The density of matter and energy throughout the Universe remains the same. This creates a balance between the Material sector and the Cosmic sector and thereby creates a cyclic Universe.

The Universe creates an infinite cycle of birth and death of stars, galaxies, and so forth, as further explained hereunder. The Universe is always changing, but on the whole remains the same. Cosmic radiation, as a result of explosions of systems in the Cosmic sector, enters our Material sector and form the vibrations that are seen as subatomic particles. These subatomic particles then form atoms, these form dust clouds in Space from which form by agglomeration red giant stars and subsequently ordinary stars. These form in agglomerate star clusters and these in turn form and feed galaxies. Stars at the end of their lives explode as (super) nova's and form pulsars, double stars and solar systems. Galaxies at the end of their lives explode and form radio systems and quasars. These explosions will be clarified later in chapter 3.4 on novae and quasars.

There is thus continuous creation of matter in the Material sector. This deviates from the view of science. Science believes that all matter was created in one instant, during the Big Bang.

And the same process as described above is happening in the Cosmic sector. These similar processes will ultimately also result in explosive products that enter as cosmic radiation into our Material sector as is reflected by the Cosmic Microwave

Background (CMB) radiation, which is everywhere and has the same value everywhere. In this way an infinite 'wheel of change' arises.

3.1.6 Summary and Conclusions

In summary, and in deviation from current scientific theory, we conclude that:

- The Universe does not expand in an accelerated manner. The apparent acceleration found by science is the result of scalar motion and incorrect interpretation of measured redshift.

- Therefore 'Dark Energy' is not needed to explain the acceleration and thus does not exist.

- The Universe did not start with a 'Big Bang'.

- The Universe is cyclic.

- Whether the Universe had a beginning and whether it has an end can not be answered by the theory.

3.2 The Material Sector

3.2.1 General

The Material sector in the view of this book is the result of the action of two antagonistic vibrations. One is the scalar in Space expanding vibration, the Progression, which pulls all the material objects in it away from each other. The other is the Gravitation, which arises from a rotational vibration on a linear vibration on the Progression, with a speed greater than the light speed, and which draws all the material objects together.
This vibration on the Progression that causes the effect of Gravitation is greater than the light speed, thus drawing all the material into itself, thereby increasing mass. If that vibration is less than the light speed, there is no Gravitation and all masses move away with the Progression and thus separate all objects from each other with the speed of light.
Cosmic radiation from the Cosmic sector eventually changes in the Material sector into the most primitive atomic structure, namely hydrogen. Because Cosmic radiation enters the Material sector uniformly, the Cosmos is filled with hydrogen atoms in the form of dust and gas. Possibly half of the total mass of the Cosmos is dust and gas.

Cosmic objects are then formed as follows.
Dust and gas form hydrogen atoms under the influence of Gravitation. These atoms will in principle separate in scalar manner from each other under the influence of the Progression. At the same time, however, they can move towards each other under the influence of Gravitation and as such form "particles".
At the centre of such a cloud of particles, the net motion is outward with the Progression, but the gravitational effect on the outer particles becomes larger with the radius of the group, because of more available mass. Somewhere a balance is reached.
Because there is a continuous influx of dust and gas from the Cosmic sector, evenly distributed throughout the entire space, the mass of such a group is increasing. As a consequence, the effect

of gravitation increases and contraction develops. Once started, it strengthens itself and is speeding up.

At the same time, in this cloud of particles, local clumping of matter occurs, that form nuclei of matter with empty Space around. These nuclei eventually become stars and the whole becomes a star cluster.

In this way, stars emerge in large round clusters, which may contain more than one million stars.

From this we can see that stars exist not in isolation from each other, although they are far apart. They take equilibrium positions among each other through the counteracting actions of Progression and Gravitation.

Each star is beyond the influence of gravitation from neighbours and therefore moves away with the Progression from all other stars. But all stars are also subject to the influence of the gravitation of the cluster as a whole.

As a result, the outer stars will move inwards and the inner stars will move outward. In result a balance will be reached. And each change means that a new equilibrium will be found.

3.2.2 Star Formation

If objects that have initially been in contact with each other separate, they will be separated by 'empty Space'.

Because of the symmetry between Space and Time, it is also possible that objects separate in Time and thus be separated by 'empty Time'.

If we look at this cosmologically for stars then we will see the following:

- A star with a lot of 'empty Space' has a large volume and low density, radiates of a large surface and thus has a relatively low temperature, which shows as a red colour. A so-called 'red giant'.

- A star with a lot of 'empty Time' has a small volume and high density, radiates from a small surface and thus has a relatively high temperature, which shows as a white or blue colour. A so-called 'white dwarf', such as Sirius B.

This explains this kind of stars for which cosmology currently has no explanation.
It is clear therefore that in the explanation as given in this book the process is just the other way around.

3.2.3 Why Stars Radiate

According to current physics stars radiate as a result of nuclear fusion, namely conversion from hydrogen to helium. However, there is no evidence for that. Because, if the energy of the star is generated by conversion of hydrogen to heavier elements, the hottest and biggest stars, such as the star Rigel, must be young. And therefore they would not live long because the amount of hydrogen would be quickly used up in this way.
But the explanation given in this book is that matter moves inwards scalar until there is only one unit of Space left. That increases the available Time units. That happens by motion as torsion, the Einstein-Cartan effect, as will be explained in chapter 3.3.4.
And the lump of matter thereby builds up thermal energy because empty Time units are energy. This happens according to the third power of the diameter (volume), while the star radiates this energy according to the square of the diameter (surface).
So, stars do not radiate as a result of nuclear fusion, but as explained above, and as such they form in their interior ever heavier elements.

Because the accumulation of matter to form a star from a cloud of dust and gas is a process that takes a very long time, it can be concluded that hot massive stars, such as white or blue stars, are ancient stars such as the Pleiades and the stars in Perseus.
Also the enormous delivery of energy such as radiated by the star Rigel is an indication that these must be old stars. Relatively cold stars are thus young stars.

We will also see later that there is a limit on the accumulation of the mass of a star, after which the star will explode. That is another reason why big stars are relatively older stars.
This process of star formation and evolution is exactly the opposite of what is proclaimed by science.

The above-mentioned process applies to all matter and therefore also to planets. Hence, also to our Earth.
That explains why planets emit more energy than they receive from the Sun, with Jupiter and Saturn (even 2.5 times as much) as clear examples, for which science also has no appropriate explanation, as evidenced by the following quote.

Quote
> Science20.com November 2010
> *That Saturn actually emits more than twice the energy it absorbs from the sun has been a science puzzle for decades but long-term data from the composite infrared spectrometer (CIRS), when combined with information about the energy coming to Saturn from the sun, could help scientists understand the nature of Saturn's internal heat source.*

Unquote

Another indication that this process applies also to our Earth is the following. Very precise astronomical measurements indicate that a day is increasing in length by about one thousandth of a second per day per century. This is because the Earth is slowing down. Science believes that this is caused by the sinking of iron from the mantle of the Earth into the core. That means some 50,000 tons of iron every second, according to calculations! In this way science assumes that the Earth was made initially with enormous amounts of iron in its exterior parts!
Clearly, our explanation of mass moving inward provides a better explanation.
The fact that we do find heavy elements in the exterior parts, even uranium, can be explained by the fact that between the core and the mantle an enormous amount of magma circulates material

between the inner and exterior parts, and can thus bring heavy elements from core to mantle.

3.2.4 Star Cluster Formation

Where the gravitational limit prevails locally, matter thus forms bigger objects. Within the gravitational limit, uniformly distributed dust and gas throughout the Cosmos combine to form stars and star clusters. Within the gravitational limit, star clusters form galaxies. These also combine and form even bigger galaxies. Galaxies are generally surrounded by hundreds of spherical star clusters. There are some 200 star clusters known around our own Milky Way. Examples are the nearest one to us which is the Sagittarius cluster at 80,000 light-years, then there is the Large Magellanic Cloud at 170,000 light-years, etc.

These star clusters form uniformly in Space and are in a distributed manner located around galaxies and are pulled by gravitation towards the centre of the galaxy system. They do not really join the rotation of the galaxy.

When they reach the galactic disc of the system, they are gradually pulled apart and the individual stars will be incorporated into the galactic structure.

In this manner, the density of a star cluster decreases the closer it comes to the galactic disc.

Star clusters with the highest density are young star clusters (M67 type) and those with the lower density the older (Perseus type).

Young star clusters have therefore the highest density, are furthest removed from the galaxy and generally consist of lighter elements. Older star clusters are closer to the galaxy, as such have a lower density and contain many heavy elements such as iron.

Galaxies then grow as a result of collisions and because of fusion with smaller systems.

The largest galaxy systems are therefore the oldest.

This explanation is contrary to the ideas of science that star clusters have just missed the formation of a galaxy and now turn around it permanently (*von Weizsäcker*).

3.2.5 Summary and Conclusions

In summary, and in deviation from current scientific theory, we conclude that:

- Star formation and evolution arise from the action of Progression and Gravitation on cosmically generated particles, thus not by condensation.

- Large massive and hot stars are old stars and not young stars.

- The explanation for a red giant is that it is a star with a lot of empty Space. A white dwarf is a star with a lot of empty Time.

- Stars radiate because their mass moves in Time and thereby convert units of Space into units of Time, which is energy. This through the Einstein-Cartan effect. Hence, not by nuclear fusion.

- Star clusters feed galaxies. They do not move around a galaxy permanently but will eventually be devoured.

- Planets form in their core heavy elements, as a result of gravitation.

- Planets also produce energy in their core, because gravitation reduces the number of Space units, increasing the number of Time units, and as such producing heat in the core.

3.3 Galaxies

3.3.1 General

Galaxies form and grow as a result of collisions with and mergers of star clusters and smaller systems, but also from collisions with other galaxies.
The largest galaxy systems are therefore the oldest.
Younger galaxy systems are elliptical, spiral galaxies are mature in the theory of this book. For science, it is the other way around.

NGC 1300 Elliptic galaxy

Our own Milky Way is spiral-shaped and orbits around as a kind of flat disc in about 200 million years. It has a cross section of more than 100,000 light-years and a thickness at the centre of over 3000 light-years. It is estimated that our galaxy contains 400,000

billion stars. Around the core, called Sagittarius A, there are 4 large arms and at least 2 small arms.
Our solar system is located in one of the smaller arms.

NGC 5457 Spiral galaxy

Galaxies can also converge. If 2 galaxies collide, it will be like a waterfall falling into a pond. There will be a small penetration of the one into the other but the effect will be more of a wall than a passage. It will be a disturbance of an existing equilibrium.
Thus, the great emptiness between the stars will not ensure an easy passage, but rather the opposite.
This is also observable in the Cosmos and the Hubble telescope has made recordings of galaxies that collide.

Located 300 million light-years away in the constellation Coma Berenices, the colliding galaxies have been nicknamed "The Mice" because of the long tails of stars and gas emanating from each galaxy. Otherwise known as NGC 4676, the pair will eventually merge into a single giant galaxy.

Also our own Milky Way will converge with another system in due time. The Andromeda galaxy (now 2 million light-years removed from us) is moving towards us and there will eventually be a merger. However, this merger is still another billion years into the future.

3.3.2 Dark Matter

According to science about 84 percent of the matter in the universe does not absorb or emit light. 'Dark matter', as it is called, cannot be seen directly, and it hasn't yet been detected by indirect

means, either. 'Dark matter' is called dark because it does not appear to interact with observable electromagnetic radiation, such as light, and so it is undetectable by existing astronomical instruments.
'Dark matter's existence and properties are inferred by science from its gravitational effects on visible matter, radiation and the structure of the Universe.
It was detected when the velocity of stars in a nearby galaxy were measured. The rotation curve of the galaxy was flat, meaning the stars in the outer spirals of the galaxy were orbiting at the same speed as stars near the centre. More alarming, the stars in the outer spirals were orbiting so fast they should have flown apart. The mass of visible stars wasn't enough to hold the galaxy together. There was an extraordinary amount of matter missing. It was called 'Dark matter'.
This shadowy substance is now thought to pervade the outskirts of galaxies, and may be composed of 'weakly interacting massive particles' or WIMPs. Worldwide, there are several detectors on the lookout for WIMPs, but so far not one has been found.
Primary evidence for 'Dark matter' comes from calculations showing that many galaxies would fly apart, or that they would not have formed or would not move as they do, if they did not contain a large amount of unseen matter.
In spite of the above, science has found that there are galaxies that apparently seem to exist with hardly any 'Dark matter' at all, such as NGC 1052-DF2. See the picture next page.

In the cosmological model of this book 'Dark matter' is not required, as the effects of Progression and Gravitation on matter create a balance, called the gravitational limit that keep a galaxy gravitationally in balance.
This balance is why galaxies would not fly apart.

Hubble Space Telescope image of NGC 1052-DF2 show lack of dark matter

The scalar Progression of the Universe is present everywhere, but a material object in it is local and locally creates a change in overall vibration. And this by the scalar character according to the quadratic law. That shows itself as the effect of Gravitation.
As long as the Gravitation is greater than the Progression, matter goes inwards towards all other Space locations and thus towards each other. That is the case locally in a galaxy.
The flat disc of a galaxy therefore operates more or less like a high viscosity fluid, as long as there is a balance between the Progression and the Gravitation.
Because gravitation depends on distance, there is a compromise between Progression and Gravitation. As mentioned before, we call this compromise the gravitational limit.
If the expanding force is greater than the gravitational limit, all galaxies are moving away from each other at high speed. Within the gravitational limit the balance will ensure cohesion in the manner as described above.

That cohesion ascertains that the speed at which stars orbit the centre of a galaxy does not decrease proportionally with the distance to the centre.
Therefore, there is no 'Dark Matter' required to achieve this effect, and 'Dark Matter' therefore does not exist.
Again, as for 'Dark Energy', we see here an example of an "auxiliary hypothesis", an ad hoc postulate that is added to a theory in response to observations that falsify it. The cosmological model of science is wrong and therefore requires such ad hoc postulate.

3.3.3 The Electric Universe

Galaxies form clusters and clusters will form superclusters. The nearest supercluster complex is called the 'Great Wall' and consists of an immense sheet of galaxies measuring about 600 million light-years in length, 200 million light-years in width, and 20 million light-years in thickness.
This cluster formation arises because throughout the whole Cosmos galaxies are connected by 'wires', which as such form a structure.
Between such superclusters there will be huge gaps, with diameters of 60-150 million light-years. The explanation for these huge gaps will be given later, because we need to deal with another characteristic first, the 'wires' of the electric Universe.
The explanation for this wire formation is as follows. As explained earlier, magnetic fields (2 dimensions) and electric currents (1 dimension) can concentrate into matter (3 dimensions).
A plasma is an ionized gas. An electric current through a plasma produces a cylindrical magnetic field which again attracts other currents that move in the same direction (compare a lightning bolt). This forms an ever-thicker circulating current that stretches the plasma and as such forms a plasma wire.
These are called Birkeland currents, after the Norwegian scientist Kristian Birkeland.

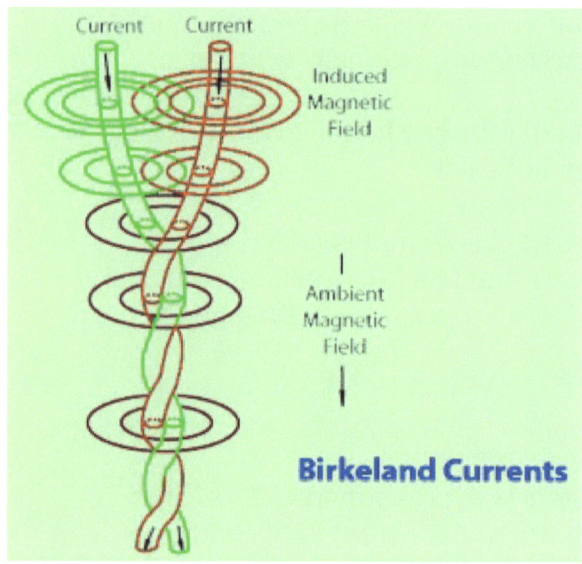

Birkeland Currents

This provides a feedback effect, because the plasma 'threads' tighten more and more with increased speed of the threads. The rotational speed increases thereby as well (similar to a pirouette, that rotates faster with the arms close to the body than with the arms extended). At the resonance points of these 'threads', these concentrate into atoms.
This process ultimately leads to the formation of galaxy systems and clusters of galaxies. Each mass absorbs and ejects scalar radiation and is thus a natural resonator.
A galaxy will rotate in the magnetic fields of the intergalactic space, thus generating electric currents.
These huge currents then flow into large wire-shaped spirals toward the centre of the galaxy, while the wires are increasingly clinging together.

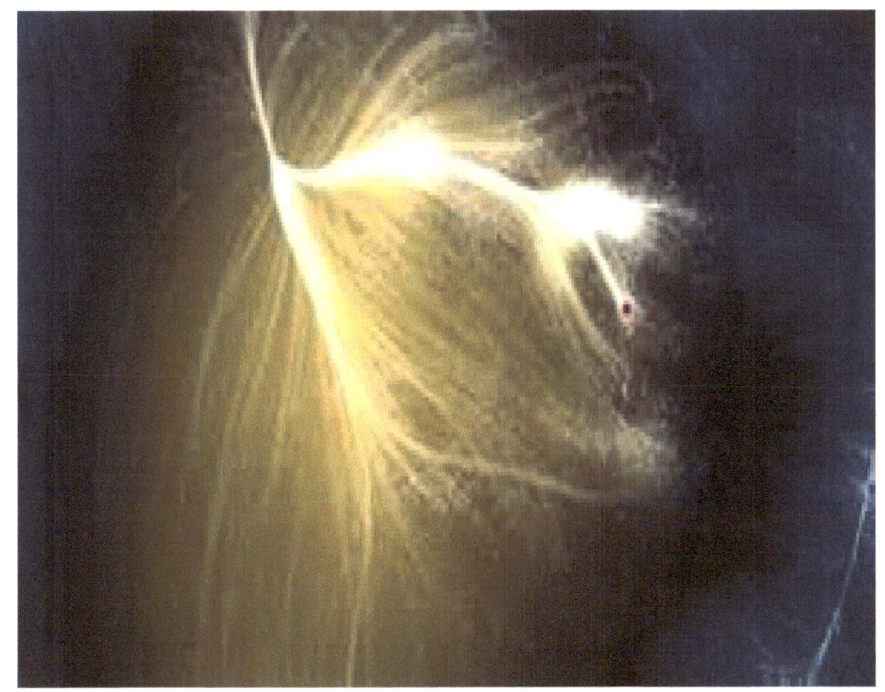

Cluster formation The red dot is our Milky Way

Space will progressively be diminished in the direction of the centre, where the current turns and along the centre axis tries to exit. The less Space means the more Time, thus Space being turned into energy.
Ultimately, there will be a 'short circuit' (no empty Space units left) as the enormous amount of energy formed will then be released explosively. This allows for powerful electric fields at the core that accelerate intense plasma jets outwards along the centre axis. An example of a galaxy with such plasma jets from the central region is M87 in the zodiac sign of Virgo.

Streaming out from the centre of M87 like a cosmic searchlight is one of nature's most amazing phenomena: a black-hole-powered jet of subatomic particles travelling at nearly the speed of light.
Credits: NASA and the Hubble Heritage Team

This electromagnetic process is active throughout the whole Cosmos, and therefore applies also to the formation of individual stars. For our sun, this explains the solar flames; these are a result also of explosively released energy. These subsequently cause magnetic storms on Earth.
As said, eventually an explosion will occur at the core.
 'Black holes' as a singularity (*Wheeler*) therefore do not exist!

The Fermi Gamma-Ray Space Telescope has discovered two huge 'gas bubbles' of about 25,000 light-years across that appear to be emitted from the centre of our Milky Way, on both sides and perpendicular to the plane of our galaxy. This shows that this finding in our own Milky Way supports the above theory.

Galaxy jets with 'gas bubbles'

3.3.4 Einstein-Cartan Effect

The explanation for the explosive release of energy lies in the Einstein-Cartan effect, whereby torsion ultimately deforms Space so that a gap is created towards Time, to the 'zero point', to the vacuum, and energy is released.
Imagine holding a can in both hands and twist it. It will result in thermal heat and the can will eventually tear.
A vortex always comes in equal but opposite pairs. The Space/Time ratio thus decreases towards the centre. And the less Space the more Time and vice versa. When squeezing down units of Space, huge amounts of energy are released.
This effect is also seen in the explosion of stars, in novae, and in quasars.
The anti-clockwise rotation of the galactic and planetary discs will result in electromagnetic torsion in the centre. Torsion fields are

formed from the spinning Sun and orbiting planets and a similar effect in a galaxy. Many rotations and long durations are involved in this process.

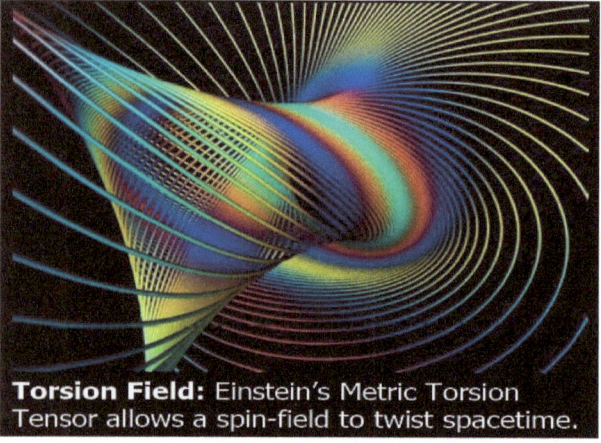

Torsion of Space/Time

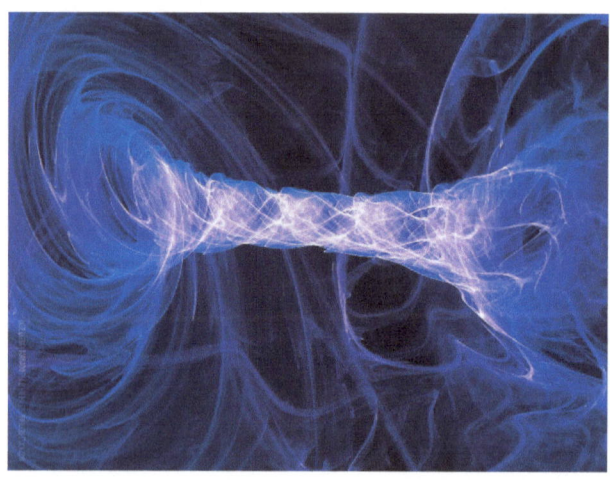

Electromagnetic torsion (laboratory)

Science is already experimenting with this torsion effect. Scientists have crafted tiny silica "dumbbells" that are too small to be seen with the naked eye, spinning them faster than any other human-made object on Earth.

On Earth, we see the effect of torsion occurring for example in tornadoes and hurricanes, whereby with the rotation also the energy content increases. The fact that tornados are often accompanied by strange green tints to the atmosphere as well as lightning on the inside provides evidence that there is an aspect of rotating electromagnetic fields involved.

This often leads to the occurrence of strange phenomena such as levitation and dematerialization, as shown in the accompanying pictures here.

These effects were first noticed by *John Hutchinson* using Tesla coils and other equipment creating complex and strong electromagnetic fields. Upon investigation by the Canadian government these effects were named Hutchison effects.

Levitation
Hurricane Andrew 1992 Florida Photo Roger Edwards

Dematerialization
Hurricane Andrew 1992 Florida Photo Roger Edwards

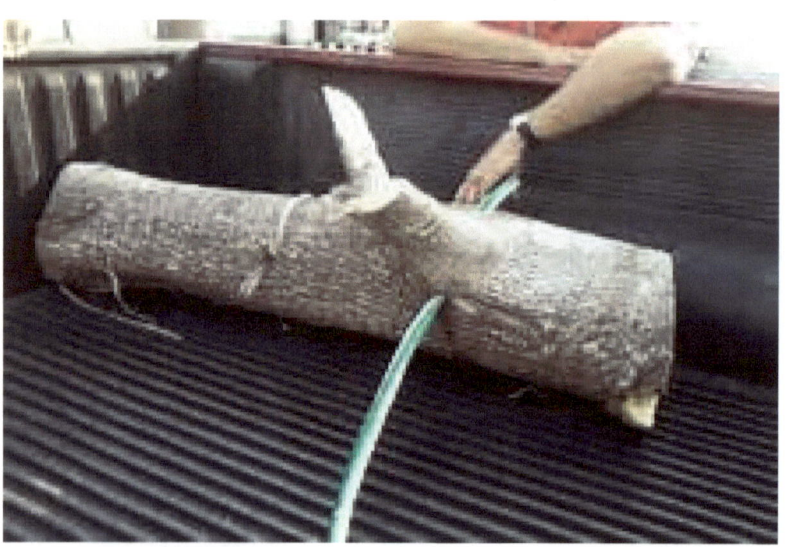

Dematerialization

Summarizing:
Matter pulls inward scalar until there is only a minimum of empty Space left, ultimately one unit of Space. In the core of the matter one finds the heaviest elements therefore. Little empty Space means many units of Time, hence a lot of energy. That energy may then be released and this can happen explosively.

> In speculation, this may also suggest an alternative theory for the theory of continental drift for our Earth.
> Pangea, the very ancient continent, was once a fully connected continent surrounded by an ocean. This continent broke up and eventually became the existing land masses (Alfred Wegener).
> It is possible that Pangea's break-up arose from a growing Earth. On a smaller Earth, larger animals could have lived, like the dinosaurs probably did, because of the lower mass. It is difficult to imagine how an animal of 85 tons (the brontosaurus) could have walked up a slight incline under present conditions of gravitation. In comparison, the largest animal today, the elephant, weighs about 7 tons.
> Maybe a sudden increase of the Earth meant the end of the dinosaurs instead of a collision with a big meteorite (?).

In support of the above looking at the planet Venus one sees a surface that is smooth and appears to be of same age, over the whole surface of the planet. It seems as if in the past a cataclysm took place that remodelled the surface
A sudden, maybe explosive increase from inside?

Explosive release of energy by conversion of Space units into Time units is the cause of dynamics in the Cosmos.
Such explosive releases are the cause of phenomena like novae and quasars.

3.3.5 Summary and Conclusions

In summary, and in deviation from current scientific theory, we conclude that:

- A galaxy operates relatively as a solid disc, due to a balance (the gravitational limit) generated by the workings of the Progression and the Gravitation.

- 'Dark Matter' therefore does not exist.

- The rotation of a galaxy and the formation of galaxy clusters are a result of electromagnetic action.

- A galaxy ultimately explodes from the centre, through the Einstein-Cartan process.

- Explosive release of energy by conversion of Space units into Time units is the cause of dynamics in the cosmos.

- Black holes as a singularity do not exist.

3.4 Novae, Pulsars and Quasars

3.4.1 Novae

A clear example of a nova is SN 2006gy.
SN 2006gy occurred in a galaxy 240 million light-years away and was the brightest and most energetic supernova ever recorded when it was discovered in 2006.

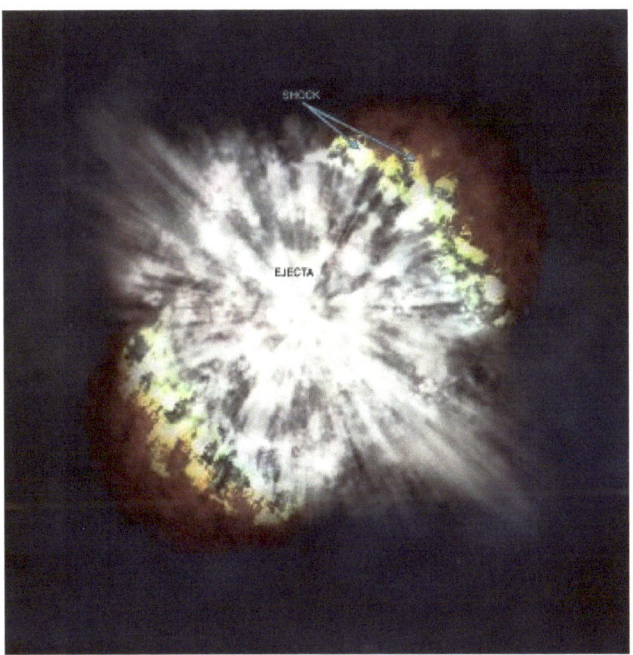

Supernova 2006gy

Science explained this phenomenon with the following theory.

Quote
> "*Initial analysis of SN 2006gy suggested that the supernova happened after a giant star ran out of fuel, with the star's core then collapsing under its own weight into an extraordinarily dense nugget in a fraction of a second and rebounding with a giant blast outward. However, such a "core-collapse" supernova likely would*

> *not have generated an iron envelope with the kind of mass and expansion rate that the new study calculated. Instead, a scenario was developed that suggests that SN 2006gy was a nova in which one star pours enough fuel onto a dead star known as a white dwarf to trigger an extraordinary nuclear explosion".*

Unquote

However, on basis of the theory developed in this book another explanation can be given that better explains this phenomenon.
A star increases in thermal energy because matter pulls inwards in scalar manner until there is ultimately only one empty Space unit left, as a result of which the available Time units increase.
This as a result of the Einstein-Cartan effect, as explained in chapter 3.3.4, where the planetary disc in counter-clockwise rotation squeezes the Space units.
Matter then builds up thermal energy because 'empty Time units' are energy. This continues until the star eventually explodes as a nova, or supernova depending on the mass, with most of the star exploding into Space.
It should be noted that this is therefore not a chemical explosion! (Astronomers suspect that the star Betelgeuze, in the constellation Orion, is now about to explode as a nova, although it may take another 10,000 years).
Because the atoms already have enormous speeds due to the increased energy, and the additional kinetic energy even adds to the speed of the explosion, the light speed can be exceeded in this way, in one or more dimensions.

A nova is thus an exploded star.
Part of the explosion product remains below the light speed and another part exceeds the light speed (in at least one dimension).
The explosion product is an expanding cloud of matter, centred on the location of the explosion.
The part that remains below the speed of light forms a cloud of dust and gas. That cloud eventually compresses into a star, but it is initially a diffuse star, a red giant. Its properties are huge size, low density and low surface temperature.
This is because there is a lot of 'empty Space' between the atoms.

The process for matter blown beyond the speed of light is the same as for the part blown below light speed, but then reversed. The explosion product expands in Time, which is thus inversely compressed in Space and thus forms a 'white dwarf'.

This inversion also applies to speed and if the force of the explosion is insufficient to lift the expansion in Time over the other two dimensions of the speed of light, the matter remains in the same spatial location and is just as visible as when it reached the speed of light. Thus, it remains a spatially identifiable object. The atoms are now separated by empty Time, rather than empty Space as is the case with the atoms of a common cloud of dust and gas from a red giant. This means that there is compression because more Time means less Space.
This creates a very small star, highly compressed with high density and high temperature, a 'white dwarf'. Due to this inverse formation process, the heaviest parts are on the outside (the iron envelope of science) and the lighter parts are on the inside.
A normal star will condense and become warmer, but due to this reverse process, a white dwarf will gradually lose energy and eventually end up under the light speed.

Together, the red giant and the white dwarf star will form a double star.
The two components of the explosion are therefore identical, both an expanding cloud of dust and gas, but where the first (red giant) expands in Space, the second (white dwarf) implodes in Time (an inverse expansion).
Initially these two stars will not be visible. The red giant does not yet generate enough heat or light to be visible (except when illuminated by nearby stars). And the white dwarf is too hot to be visible, because the radiation is transmitted mainly in the X-ray band. From here the Gravitation takes over which causes the red giant to heat up and the white dwarf to cool down and the radiation of both now come within the visible part of the spectrum. This process continues until the next nova of stars.

Most stars in our galaxy are such double stars and in some cases even more than double. For example, Castor in the constellation Gemini consists of at least 6 components.
Science has problems to explain double stars because they assign different ages to the two stars by their appearance, while they were clearly created at the same time.

There are several reasons for suspecting that our Sun is part of a double star system too and much research is being done in this regard, for example by the BRI (Binary Research Institute).
Should that be the case, than this might also explain the precession of the equinoxes. That would then explain the wobble in the axis of the Earth, which causes the place where the Sun rises to backtrack one star constellation of the zodiac every 2,000 years.

In a supernova explosion the same happens, and a white dwarf is also created here. But now the power of the explosion can be large enough to blow some of the explosion product over the light speed into the Cosmic sector. The speed of the explosion product now moves over 2 or possibly even 3 dimensions of the light speed.
That means leaving the Material sector and entering the Cosmic sector.
In this Cosmic sector, the explosion product also consists of a cloud of dust and gas. The atoms will once again form Cosmic stars, but now expanded in Time.
Following that, Cosmic galaxies are formed and eventually the same processes take place as in the Material sector. That is, eventually exploding and returning in the form of Cosmic radiation into the Material sector.

The following is a picture by the Hubble space telescope of the Calabash nebula.

The rotting eggs nebula

The Calabash nebula is a nova explosion of a star with about the mass of our Sun at about 5000 light-years from us.
The star explodes clearly in opposite directions and this at huge speeds.
The left part of the picture is clearly less stretched than the right part, while the explosive force is the same, which indicates that the left part has exceeded the light speed. Therefore, part of the energy is needed to overcome gravitation in Time and therefore that part of the explosion remains located in Space.
The whole will transform into roughly 1000 years into a nebula, now known as the nebula of rotting eggs due to the high concentration of sulphur.
The left part will eventually form a white dwarf star and the right part a red giant.

3.4.2 Pulsars

A Pulsar is a pulsating radio source in Space. Pulsars were first discovered only in 1967. This radio radiation is observed on Earth in the form of fast pulses, as if it were a lighthouse in Space. Pulsars produce radio radiation with a precision like the radiation of a laser. Pulsars send such radio radiation pulses into Space with a precision such that they approach the accuracy of atomic clocks. And these pulses have very short periods, of milliseconds to a second! In addition, they are very bright.
Since its discovery, cosmology catalogues pulsars.
In the Crab Nebula sits a pulsar that is unique among the pulsars, as this pulsar emits pulsating radiation of frequencies covering the whole spectrum, including the frequency ranges of radio, optical, X-ray and gamma-ray radiation.

Science explains this phenomenon by the theory that pulsars are extremely fast-spinning stars. That must then be extremely small stars because the peripheral speed for such a fast pulse would otherwise be impossible without the star flying apart. But if such a star is to be very small, it creates another problem again for science to explain the brightness.

The explanation given here for a pulsar is that a pulsar is an exploded star (nova), with some of the explosive material moving faster than the speed of light. Only that part of the explosion speed needed to overcome Gravitation has an effect on speed in Space. At speeds greater than the light speed, part of the explosion speed is used to overcome Gravitation and the other part of the speed goes in Time and thus opposite. That means that the faster a nova goes, the less distance it covers.
The same applies to the radiation. Part of the radiation remains visible in Space, the other part is in Time. There are no fractional units possible because a unit of Space and a unit of Time are the minimum units. Thus, during a unit of Time, the radiation must be either Space directed or non-Space directed. A speed under a Space/Time unit is thus not possible, but a speed under unit speed can be produced by adding units of energy (Time). Thus, the

reduction in spatial radiation, which is below the level of one unit of radiation per unit of Time, is processed as the number of units of Time during which the radiation is spatial. That means fluctuating. That explains the phenomenon of pulsar, for which science does not actually have a satisfactory explanation.

3.4.3 Quasars

A Quasar (Quasi Stellar Object) is a Cosmic phenomenon of enormous luminosity, but also with a huge red shift in the spectrum. Normally, a maximum of 0.5 is measured for a galaxy, while for quasars more than 2 can be measured. That is why science has assumed that quasars are far removed from us, even billions of light-years. But because of the great luminosity of a quasar in conjunction with the vast distance, it becomes an inexplicable phenomenon for science.

Quasars, however, are not far away from us. Quasars are usually observed in connection with so-called Seyfert galaxy systems which have less Doppler values than the quasar (and are thus closer by). Therefore, a connection between those two can cosmologically be expected.
(Refer *Dr. Halton C. Arp* Atlas or Peculiar Galaxies).

In each galaxy, the stars towards the centre are the oldest. They will therefore regularly explode as a nova. Therefore, the core of a galaxy will emit radiation into the radio frequency band. With increasing age of the system, such explosions will increase and the core will grow and the pressure in the core will thus also increase. This results in what is called a Seyfert galaxy. Eventually, the pressure becomes so large that it breaks through the overlying material in an explosion.

Seyfert galaxy systems comprise roughly 10% of all galaxy systems and have super massive nuclei. They look like normal spiral systems, but the cores often radiate as much as our entire galaxy.

And according to *Halton Arp*, Seyfert galaxy systems that show a low red shift in the spectrum are often clearly linked to galaxy systems with a very high red shift in the spectrum.

An example is the galaxy NGC 4319 and the nearby Quasar Markarian 205, with different red shifts but which still appear to be connected. *Dr Halton Arp* was the first to observe such a connection.
Also later photographs indicate a bridge between the systems. See, for example, *http://davidpratt.info/astro/arp5.jpg*
However, a definitive answer has not yet been given.

Galaxy NGC 4319 and Quasar Markarian 205

Hereafter a picture of galaxy NGC 7319 (Doppler-value 0.0225). The small object at the arrow is a quasar (Doppler-value 2.11). This observation of a quasar between the Earth and the galaxy shows that the quasar cannot be further away than the galaxy.

Galaxy NGC 7319.
Arrow indicates a quasar in the foreground

So, what causes such a high red shift in the spectrum? The explanation for this is as follows.
In a quasar, a galaxy explodes from the core, the 'black hole'. It basically consists of fallen stars and thus the explosion product consists of a cloud of stars, in other words small galaxies.
The explosion process is actually identical to what we have seen in a nova and supernova.
The explosion occurs as a result of the Einstein-Cartan effect, where the disc of stars in the galactic arms in counter-clockwise rotation squeezes the Space units.
In a nova, the part that explodes under the light speed forms a cloud of atoms and molecules, and the part that explodes at a speed greater than the light speed becomes a white dwarf star.

Similarly, in an explosion of a galaxy the explosion product that remains below the light speed forms a cloud of stars, thus a small galaxy. It initially emits a strong radio radiation and is therefore called a radio system.

The other part that exceeds the light speed forms a quasar. The quasar is then in Space/Time, so it looks small but radiates tremendously. A quasar is actually a 'white dwarf' galaxy.

In explosions of such enormous masses, the explosion products achieve huge speeds. The material is then on both sides of the Time-Space and Space/Time boundary. This manifests itself on the Time-Space side as the 'jets' on both sides of a galaxy.

The bulk of matter disappears into the Cosmic sector. Because the quasar goes faster than light, it expands in Time and thus shrinks in equivalent Space. As a result, a quasar appears to be small with high radiation.

In a quasar, a substantial part of the mass of a galaxy is converted to energy, but because the light speed has been exceeded, the whole disappears from our Material sector of the Universe.

Because the explosion takes place in the Cosmic sector the effect will resemble radioactivity. So from our viewpoint the explosion will go slowly and the product will remain visible for a long time.

A red shift of 1 corresponds to the Progression, so with the speed of light. Above this the quasar becomes smaller for our perception with increasing red shift. Theoretically, a quasar remains visible to a red shift of 2.3. All quasars found to date have a red shift that is smaller than this theoretical value, indicating that the theory is supported by observation.

After that, the light speed has been exceeded in 3 dimensions and the quasar becomes invisible because the quasar has now completely disappeared into the Cosmic sector.

Only at the location a rest of dust and radiation may remain visible for a while. Matter that not yet has managed to make the full transfer. That will than look like a glowing cloud of dust, of which the cause of the glow is then a non-identifiable source.

A recent example is given hereunder, from an article in NU.nl/Dennis Rijnvis:
"*Astronomers discover gas cloud with mysterious glow*",

published: 25 February 2017 and copied from scientific journal *Astrophysical Journal*.

Quote
> Scientists have discovered a gas cloud with an inexplicable glow at 10 billion light-years of the Earth. It is a cloud of glowing gas in a cluster of thousands of galaxies covering an area of 50 million light-years.
> A large star or other light source cannot be seen near the gas cloud. It is therefore unclear why the cloud radiates light.
> The newly discovered gas cloud belongs to the 'huge Lyman alpha nebulae' a category of very rare space objects. These are large gas clouds formed in the space between different galaxies. The newly discovered cloud has been named MAMMOTH-1. According to the main investigator Xavier Prochaska, the light of the gas cloud is particularly well visible. "It's extremely fierce, we're talking about an incredibly strong kind of energy without visible source," he explains in the International Business Times.
> The scientists suspect that the cloud owes its special glow due to a quasar, a clear nucleus of a far removed active galaxy.
> This light source cannot be seen on the telescope images. "But I suspect it is largely hidden from the quasar by dust, so only part of the light can be seen," said Prochaska.

Unquote

The astronomer suspects that the source (a quasar) is being extracted from view by dust. However, according to the theory described here, the quasar has completely gone through all three dimensions of the light speed fully into the Cosmic sector and is therefore no longer visible. Also note that there is mention of large gas clouds formed in space *between different galaxies*, so in a place where a galaxy might have existed.

The explosion in a quasar is not caused by thermal energy like in a nova, but in this case by ageing, by taking in an overdose of neutrinos. As mentioned before, neutrinos usually pass through matter, but sometimes they get a charge and then cannot exit any more. In the long run this produces ageing.

The quasar phenomenon provides therefore also an explanation for the question of why information is not lost, an aspect that also

troubles science with their theory of the existence of 'Black Holes' where information disappears in the black hole and cannot get out. The physicist Stephen Hawking, provided a theoretical argument for retention of information by 'Black Holes' in 1974.
In the definition of Wikipedia Hawking radiation is:
"Black-body radiation released by black holes, due to quantum effects near the black hole event horizon. It reduces the mass and rotational energy of black holes and is therefore black hole evaporation. Because of this, black holes are expected to shrink and ultimately vanish".

The quasar is in our theory responsible for retention of information. This because matter (energy and thus information) is moved from the Material sector to the Cosmic sector.

3.4.4 Fast Radio Bursts (FRB)

About ten years ago, scientists first discovered a fast radio burst (FRB), or a fast radio flash. Thes are flashes of light, unleashing in a few milliseconds as much energy as our Sun does in a 100 years. These bursts seem to originate in galaxies billions of light years away, as evidenced by new research. More than a 100 FRBs have been discovered to date.

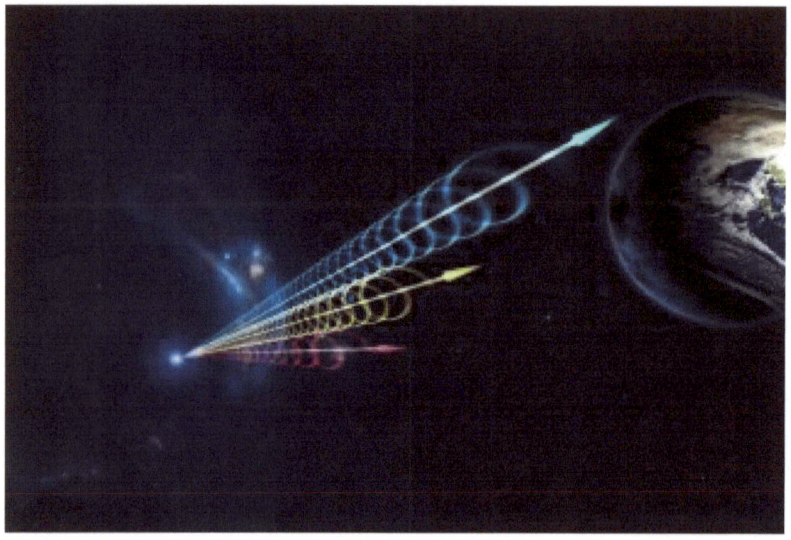

To present the view of science on FRBs find herewith an article, published: 5 January 2017 in Scientias magazine, entitled:

"Astronomers can finally bring home mysterious fast radio bursts".

Quote
> *Astronomers have for the first time pinpointed the location of a so-called 'fast radio burst' - a type of short-duration radio flash of enigmatic origin - and have used this to identify its host galaxy. The results have appeared in Nature and the Astrophysical Journal Letters.*
>
> *How exactly a FRB is created is unclear to date. Some researchers initially believed that such a **fast radio flash** occurred during a collision of neutron stars. And so, these researchers automatically assumed that such a radio flash was a one-off event. That theory became unsettled however when astronomers stumbled upon FRB 121102. This radio flash showed itself several times.*
>
> *And now the research on FRB 121102 is again a key: scientists have determined the exact location of the fast radio flash. And guess what? The radio flash originates in a **dwarf star system** at about **three billion light-years distance**. We can now say with 100 percent certainty that it is an extragalactic phenomenon. "We are the first to show that this is a cosmological phenomenon," says investigator Casey Law. "It's not something in our backyard. And we are the first to see where it originates: **in a small galaxy.** And that is very surprising in my view. "Researcher Shriharsh Tendulkar agrees with that: "One would expect the fastest radio flashes to arise in large galaxies with the largest number of stars and neutron stars (residues of heavy stars."*
>
> *How the radio flashes precisely arise, continues to be a guess for now. Perhaps they are from the collapsed remnant of a heavy star. Another option is that they are generated in the vicinity of a black hole that steals gas from its environment.*

Unquote

To explain the phenomenon of a FRB in the light of the theory in this book it is proposed that it is a quasar which has exceeded the light speed in at least 1 dimension and contains pulsars.

Note that in the article the location is found in a faraway dwarf star system, in a small galaxy, showing high energy, high frequency radio flashes.
These are all clear phenomena of quasars and pulsars according to the theory of this book.

Hence, FRBs are quasars containing pulsars.

3.4.5 Summary and Conclusions

In summary, and in deviation from current scientific theory, we conclude that:

3.4.5.1   Novae

- A nova is an exploded star, where part of the explosion product remains below the light speed and another part exceeds the light speed (in at least one dimension).

- The part that remains below the speed of light expands in Space and forms a diffuse star, a red giant. Its properties are huge size, low density and low surface temperature.

- The part that is blown beyond the speed of light expands in Time, and forms a 'white dwarf', highly compressed with high density and high temperature, in an inverse process with the heaviest parts on the outside.

- A normal star will condense and become warmer, but due to this reverse process, a white dwarf will gradually lose energy and eventually end up under the light speed.

- In a supernova explosion the power of the explosion can be large enough to blow some of the explosion product over 2 or possibly even 3 dimensions of the light speed into the Cosmic sector. In the Cosmic sector the atoms of the explosion product will form stars and later galaxies, but here expanded in Time (is spread out in Space).

3.4.5.2   Pulsars

- Pulsars are parts of a nova that has exceeded the speed of light. It generates thereby fractional units of radiation, which then manifest themselves alternately in Space and in Time. Pulsars are therefore not necessarily fast spinning stars.

- At speeds greater than the light speed, part of the explosion speed is used to overcome Gravitation and the other part of the speed goes in Time and thus opposite. That means that the faster a nova goes, the less distance it covers.

3.4.5.3 Quasars

- The oldest and largest galaxy systems eventually explode. The explosion is not caused by thermal energy like in a nova, but in this case by ageing, by taking in an overdose of neutrinos. This produces a quasar.

- In a quasar, a substantial part of the mass of a galaxy is converted to energy, but because the light speed has been exceeded, the whole disappears from our Material sector of the Universe. Because the explosion takes place in the Cosmic sector the effect will resemble radioactivity. So, from our viewpoint the explosion will go slowly and the product will remain visible for a long time. Because the quasar goes faster than light, it expands in Time and thus shrinks in equivalent Space. As a result, a quasar appears to be small with high radiation.

- With the explanation as given in this chapter the quasar becomes responsible for retention of information. This because matter (energy and thus information) is moved from the Material sector to the Cosmic sector.

3.4.5.4 FRBs

- An FRB is a quasar with pulsars in it.

3.5 The Genesis of the Universe

3.5.1 The Cyclic Universe

The Universe has an elegant, symmetrical and balanced structure, with its intertwined Material sector and inverse Cosmic sector. Both based on quanta Space per quanta Time, expressed as speed, with light speed as the basis.
The Progression in the Material sector moves galaxies away from each other at light speed into infinite Space. Gravitation in the Material sector brings matter together towards infinite Time.

There is a limit to the age and size of stars and galaxies. For stars it is a thermal limit and for galaxies it is ageing.
The limit for a star is close to 100 solar masses and the limit for a galaxy is in the vicinity of 10,000,000,000,000 solar masses. When this limit is reached, an explosion occurs which causes (a part of) the matter to end up in the Cosmic sector. There the same processes take place as in the Material sector but in reverse direction. Here too, ultimately a limit is reached upon which matter is returned into the Material sector again.

In the Cosmos, enormous areas of empty space are found that can be cross-sectioned from 60 to 150 million light-years. These are the remains of galactic explosions where the explosion product has been swept into the Cosmic sector at more than 3 times the speed of light. Matter in the Cosmic sector is spread out in Time and invisible from our Material sector. Cosmic galaxy systems also have mass but here a mass m is instead a mass $1/m$, as such still a positive amount.
This empty space has a decisive influence on matter around it. Our Milky Way and the nearby Andromeda galaxy are pushed away by such empty space for which science has no explanation but which they have called the 'Dipole Repeller'.

Matter from the remains of stars and galaxies that have reached the end of their lives is reused. This process is equivalent in both sectors of the Universe.

When the exploded matter in the Material sector has speeds below light speed, matter remains in the Material sector. When the exploded matter has speeds above light speed matter moves into the Cosmic sector. In the Cosmic sector a similar process takes place and the explosion product that reaches the Material sector does so in the form of the cosmic background radiation. Matter that remains in the Cosmic sector is anti-matter.
This process results in a balance between the Material sector and the Cosmic sector and thereby creates a cyclic Universe.

The Universe is thus always changing, but on the whole remains the same.
This creates an infinite cycle of birth and death of stars, galaxies, and so forth. Cosmic radiation, as a result of cosmic explosions of cosmic systems, enters our Material sector and forms the subatomic particles. These then form atoms, these form dust clouds in Space that form by agglomeration red giant stars and subsequently ordinary stars. These form in agglomerate star clusters and these in turn form and feed galaxies.
Stars at the end of their lives explode as (super) nova's and form pulsars, double stars and solar systems.
Galaxies at the end of their lives explode and form radio systems and quasars.
In this way an infinite 'wheel of change' arises.

3.5.2 Physical Constants

Scientists still wonder why the Universe is so precisely adjusted that certain constants in the laws of nature have values that, by the slightest deviation, would not have created a Universe or a totally different Universe.
The fact that the Universe is cyclic in our theory can provide an alternative answer to this question why the Universe is so precisely adjusted.
We here review some of these constants that science considers significant. As can be seen all those physical constants disappear in the theory of this book.

The relationship between electromagnetism and gravitation
Science: The main difference between gravity and electromagnetism is that gravity is a force between masses whereas electromagnetism is a force between charges.
We: Electromagnetism and gravitation are linked together, as a consequence of the number of dimensions in which they exist. (As $1/c$ and $1/c^2$ respectively).

The Cosmological Constant Λ
Science: The cosmological constant (lambda) is the energy density of space. The Cosmos is measured as expanding at a rate of some 70 km/sec/Mpc. Hence, a driving force termed Dark Energy has been postulated to account for this force.
We: Dark Energy does not exist. There is no acceleration of the galaxies in the Universe but there is a constant movement, called the Progression, with the speed of light. The measurements made by science on basis of redshift are incorrect.

The strong nuclear force E
Science: The strong nuclear force (epsilon) holds nucleons together. Nucleons must be extremely close together so that an exchange of mesons can occur. Mesons are the carrier of the force. The distance required is about the diameter of a proton.
We: If two atoms move together under the influence of Gravitation, this can happen only until there is 1 unit of Space left. To get even closer together, a vibration is needed in Time. This vibration works in Space inwards and in Time outwards. This provides balance because these vibrations work jointly and opposed.

The amount of matter in the Universe Ω
Science: The universe is constrained by a law of conservation of matter and energy. The majority of mass in the Universe is presently undetectable, hence its being

	termed Dark Matter. It is known to exist only because of the gravitation in the Universe, and is thought to account for approximately 85% of the matter in the Universe.
We:	Dark Matter does not exist. Galaxies are subject to the effects of Progression and Gravitation. A balance between these motions will form, called the gravitational limit. Within the gravitational limit the balance will ensure cohesion in the manner as described above. No Dark Matter postulate required.

The relationship between protons and electrons
Science:	A proton is a positively charged particle in the nucleus of an atom. An electron is a negatively charged particle that orbits a nucleus. Removal of the electron ionizes the atom.
We:	Ionization is a modification of the fundamental rotation of an atom, towards Space or towards Time.

Primordial Ripples in the Universe Q
Science:	The Cosmic Microwave Background (CMB) is a remnant from the Big Bang.
We:	The Universe is cyclic and the CMB is radiation of Cosmic objects, from objects that in the Cosmic sector are spatially distributed. As such, the radiation of the Cosmic sector is not localized, and we see that cosmic radiation in the Material sector as of low strength and isotropic. Q does not exist.

Chapter 4   Epilogue

Study the art of science and study the science of art
Develop your senses and learn how to see
Leonardo Da Vinci

The Universe in essence does not consist of matter, solid matter. Reality exists as a vast ocean of vibrations. A base linear scalar vibration of Space/Time in 3 dimensions (the Progression). On this base vibration exist scalar rotational vibrations in 3 dimensions. Different speeds are possible in each dimension, with as base speed light speed, one unit Space per one unit Time in each dimension. This creates two intertwined sectors, where speeds are below or above light speed, called here the Material sector and the Cosmic sector respectively.
We believe we understand Space. We also need to understand and be able to use Time. Time not only as time, but as energy and as information.
Thus space contains energy. This energy is what is called the vacuum energy. A perpetuum mobile is not possible in a closed system, but in principle it is possible in an open system, where the extra energy can be taken from an outside source. In the model as described in this book the Universe is such an open system, with interaction between Material sector and Cosmic sector. The vacuum energy can thus be released and made available for use by humanity. This process is not yet suitably engineered, but is in principle possible through torsion. Torsion can break Space down to minimum quanta. *"The less Space the more Time"*, means that squeezing Space will increase the amount of available energy. How to do that economically is the question that will eventually solve the energy problem of the world.

Our Universe is a system of rotating systems in other rotating systems. This creates a continuous changing system in Space and Time. Space and Time are both dynamic in this. All systems are therefore Space/Time machines, which constantly intertwine and influence each other.

The Universe is thus cyclic. The ultimate question of how the Universe came into being or where it eventually will go, cannot be answered and our theory is therefore neutral as to whether the Universe was created or has always existed. We do not know if there is a beginning or an end.

The cycles include transformation processes such as in scientific terms, transformation of radiation into a material particle, a material particle into another particle, and accumulation of particles. In our theory transformation of S/T vibrations in the Material sector into S/T vibrations in the Cosmic sector and vice versa. In this way, the Universe is comprehensible in all its details, is logical and ordered.

This concept answers some philosophical questions, such as whether no Space and/or no Time exist, and whether infinite Space and infinite Time exist.

The model simply shows that these are two aspects of the same issue, namely infinite Space equals zero Time and infinite Time equals zero Space.

Unmanifested Space/Time is the equivalent of nothing, and when we talk about the Universe we actually talk about the area where matter exists. And matter is limited in Time and is then recycled through the Cosmic sector.

As such, we can say that the Universe is finite within an infinite Space/Time cycle.

According to the second law of thermodynamics, the entropy of an isolated system that is not in balance increases over time.

In other words, in the Material sector all structures will move from a state of larger organization to a state of lesser organization. The available energy decreases.

Conversely, in the Cosmic sector, the reverse is the case and the energy increases. The Cosmic sector can be viewed as an information field, that the Swiss psychologist *Carl Jung* labelled as the collective unconscious, in which ideas are present as archetypes. *Ervin Laszlo* recognized this as the Akashic Field, and *Fritjof Capra* and *Lynne McTaggart* as The Field. Here thoughts attract creative energy as ideas. It is imagination that brings creative energy into form. This is specifically the field of art.

In the Material sector, we see that living beings can increase the level of organization. Therefore, one may deduce that:

"*Living creatures in the Material sector are connected to and controlled by an operating system in the Cosmic sector*".

And like a fractal carries the whole in itself and repeats itself at higher levels, the whole is reflected in the small. So we can see ourselves as a smaller version of the whole of reality. So above, so below (*Hermes Trismegistus*).
Such a complex interconnecting system of rotating systems produces 'knots' in Space/Time through resonance. Resonance creates the objects that we see in reality, including ourselves. Ultimately it is our brain that translates all this into what we see as the world around us.

Which brings me to a last philosophical recognition, namely that all the above may provide a good explanation for the fabric of the Universe, but fails to address the why of the existence of the Universe and also leaves out clarification of aspects such as the origin of life, emotions, and specifically consciousness.
Thus there is more to the Universe than just the fabric!
Reality is a state of awareness, which we try to understand with consciousness. This means that consciousness is the first and last principle.

www.ingramcontent.com/pod-product-compliance
Lightning Source LLC
Chambersburg PA
CBHW040314220526
45473CB00009B/2427